D0787112

SPACE: ITS IMPACT ON MAN AND SOCIETY

SPACE: ITS IMPACT ON MAN AND SOCIETY

Edited by LILLIAN LEVY

Essay Index Reprint Series

BOOKS FOR LIBRARIES PRESS
FREEPORT, NEW YORK

TO AARON, ESTHER AND STEVEN;
ALL THAT IS, HAS BEEN OR WILL BE

Reprinted 1973 by arrangement with Lillian Levy
and Maxwell Aley Associates, Inc., Literary Agents

Library of Congress Cataloging in Publication Data

Levy, Lillian, ed.
Space, its impact on man and society.

(Essay index reprint series)
Reprint of the 1965 ed.
1. Astronautics and civilization--Addresses, essays, lectures. I. Title.
[CB440.L47 1973] 301.24'3 72-13181
ISBN 0-8369-8164-2

PRINTED IN THE UNITED STATES OF AMERICA

CONTENTS

PREFACE

The years that have passed since Sputnik I was launched in 1957 have been filled with many remarkable achievements in outer space by both the Soviet Union and the United States, including man's first journey into this new frontier. Most of us are generally aware of the technological advances that have made our progress into space possible. But we have been slow to recognize the many subtle changes, the pervasive influences, that the Space Age has wrought in our daily lives—personal and professional, domestic and political, material and spiritual.

Within the first few years of this new era, astronautics has replaced aviation as America's leading industry. It has spawned new jobs and new communities; new products for consumer use have evolved from space research. Education and, indeed, the educated man have assumed new and important status; science itself is receiving unprecedented appreciation and attention. Leaders in government, industry, and education are urging greater emphasis on science and mathematics beginning in the elementary grades. The scientist is now a member of a new elite whose counsel is sought at the highest levels in our society. These changes and influences we can, to a large degree, understand and accept.

Space exploration, however, has posed serious political, economic, social, religious, and philosophical questions which demand our attention. This book is an attempt to examine these questions so that we may be better able to assess the substantive impact of the Space Age on mankind now and in the future, and to recognize the hazards as well as the high hopes of this new era. It offers for reflection and consideration the views of persons of outstanding ability and recognized authority, and I am very grateful to these distinguished contributors for the active interest and generous support they have given to this endeavor.

I am also grateful to Richard France, my friend and colleague, who encouraged me to undertake this book; and thanks are due to Ruth Aley, my agent, and to Burton L. Beals, my editor. I should also like to acknowledge the assistance provided by Dr. Eugene M. Emme, chief historian of the National Aeronautics and Space Administration, his associate Dr. Frank W. Anderson, and members of their staff. I wish to express my deep appreciation to Lois Middleton for the time, effort and literary ability she has generously given to the production of the book. Above all, I am most grateful to my husband, Aaron, not only for his patience and encouragement, but for his invaluable editorial and critical judgment.

Lillian Levy

THE CONTRIBUTORS

President Lyndon B. Johnson was the first of our national leaders to recognize the importance of space exploration as an international as well as a national goal. His concept of space as a new forum for international cooperation is reflected by the provisions of the Space Act, of which he was the chief architect. He became the first chairman of the Senate Committee on Aeronautical and Space Sciences and, as Vice President of the United States, was chairman of the National Aeronautics and Space Council. As the leading statesman of the Space Age, he has made clear his intention that this new domain be used only for the service of mankind.

James E. Webb was appointed Administrator to the National Aeronautics and Space Administration in 1961 by the late President John F. Kennedy. He brought to his NASA position a rich background in public service, law, aviation, business and education; at the time of his appointment, he was president and trustee of Educational Services, Incorporated, of the Massachusetts Institute of Technology, a non-profit organization devoted to research and improvement in the teaching of physics in secondary schools. Mr. Webb was Director of the Bureau of the Budget under President Harry S. Truman and later became Undersecretary of State in his Administration.

Joseph A. Beirne is President of the Communications Workers of America, a union that has—largely because of his foresight and initiative—grown with automation. A vice president of the AFL-CIO, he has served for many years as chairman of its Community Services Committee and as a member of a number of special committees of the Executive Council. Mr. Beirne has effectively promoted greater understanding between management and unions and has encouraged workers to undertake the education necessary for employment in the Space Age. He is the author of several articles and a book, *New Horizons for American Labor.*

Stanley H. Ruttenberg is an internationally known economist and authority on labor. Now special assistant to the Secretary of Labor, he was formerly director of the Department of Research for the AFL-CIO, a post he had held since December, 1955. Mr. Ruttenberg has served as an adviser and consultant to many international organizations and commissions. He is also an active member of the American Economic Association, the American Statistical Association, and the Industrial Relations Research Association.

Martin Goland, a recognized authority on aircraft design, applied mechanics, and operations research, is President and Director of the Southwest Research Institute in San Antonio, Texas. He is also a member of several national scientific advisory groups, including the committee on aircraft structures of NASA, of which he is chairman, and the Committee on Science and Astronautics of the House of Representatives. He is editorial advisor for the monthly publication, *Applied Mechanics Reviews,* and the author of over sixty papers on subjects which include aerodynamics, dynamics, structures, mathematics, engineering analysis and research administration.

William C. Foster is the Director of the United States Arms Control and Disarmament Agency. A former industrialist with a long history of public service, he was Under Secretary of Commerce from 1946 to 1948, and was then appointed deputy U.S. special representative in Europe for the Economic Cooperation Administration (Marshall Plan). He became ECA Administrator in 1950, and from 1951 to 1953 was Deputy Secretary of Defense. A graduate of the Massachusetts Institute of Technology, Mr. Foster has received honorary doctorates from both Syracuse University and George Washington University.

General Bernard A. Schriever is Commander of the United States Air Force Systems Command (AFSC). A graduate of Texas A & M, Stanford University, and the National War College, he was given command of the Air Force Ballistics Missile Division in 1954 and directed the nation's highest-priority project—the development of the intercontinental ballistic missile. General Schriever assumed his present command in 1961 and was promoted to four-star rank. AFSC, under General Schriever, is responsible for the research, development, procurement, and production required to place a complete aerospace system in operational use.

Nicholas deB. Katzenbach is Deputy Attorney General in the Department of Justice. A graduate of Princeton, he received his LL.B degree *cum laude* from Yale Law School, and was a Rhodes Scholar at Oxford University. He is a recognized authority on the international legal aspects of space and is co-author of a study entitled "Legal Literature of Air Space" and author of a book, *The Political Foundations of International Law*. His activities on behalf of the Department of Justice on cases involving civil rights are well known. Before assuming his present position, Mr. Katzenbach was a Professor of Law at the University of Chicago School of Law.

Dr. Glenn T. Seaborg, a Nobel prizewinner in chemistry, is the Chairman of the United States Atomic Energy Commission. One of the world's outstanding creative scientists and teachers, he was Chancellor of the University of California at Berkeley from 1958 until his appointment to the AEC by President Kennedy in January 1961. Dr. Seaborg was a co-discoverer in 1940 of element 94, plutonium, the first of more than ten transuranium elements he was to help discover in the next eighteen years. In addition to many other honors, Dr. Seaborg was named recipient of the AEC's Enrico Fermi Award for his work on nuclear chemistry and his leadership in scientific educational affairs.

R. W. Retterer is Senior Vice President of Marketing for the UNIVAC Division of the Sperry Rand Corporation. His position in the business and financial world has enabled him to observe the broad spectrum and changing character of careers and career opportunities as they have developed in the Space Age. Prior to joining UNIVAC, Mr. Retterer was associated with Waddell & Reed, Inc., and IBM (International Business Machines Corporation). In his present position he is responsible for all commercial, federal government, and defense marketing activities for the UNIVAC division in the continental United States. He is a graduate of the University of Indiana.

John H. Glenn, Jr., the first American to orbit the earth, made his pioneer space flight on February 20, 1962, and added four hours and fifty-six minutes of space flying to his more than 5,000 hours of flying time. Col. Glenn began his career in aviation as a Naval cadet in 1942. He flew a total of fifty-nine combat missions in World War II and sixty-three missions in the Korean conflict. He has been awarded the Distinguished Flying Cross on five occasions and he holds the Air Medal with eighteen clusters for his service during wartime. Since orbiting in space, Col. Glenn has served as an unofficial Ambassador of Goodwill in almost every part of the world.

Lillian Levy, the contributing editor of this volume, is a free-lance writer and journalist, now with the Office of Public Affairs of the National Aeronautics and Space Administration. As Bureau Chief for the *National Jewish Post and Opinion* and as a former correspondent for Science Service, Inc., she has reported on events at the Capitol, the State Department, the Pentagon, NASA, AEC, and the White House. Mrs. Levy has written for the *Saturday Review, Science,* the North American Newspaper Alliance, and many other publications. She is also the author of several radio scripts on science beamed abroad by the Voice of America.

Hubertus Strughold, the "Father of Space Medicine" and the man who conceived the world's first space-cabin simulator, is the first professor of space medicine at the United States Air Force School of Aerospace Medicine and is Chief Scientist of the Air Force Systems Command's Aerospace Medical Division at Brooks Air Force Base, near San Antonio, Texas. He is the author of more than 170 professional papers on physiology, aviation, and space medicine, and of the book *The Green and Red Planet: A Physiological Study of the Possibility of Life on Mars.* He is also co-author of a textbook, *Principles and Practices of Aviation Medicine.* Dr. Strughold is a member of several national and international scientific and medical organizations, and has received numerous awards for his pioneer space research.

S. Fred Singer is one of this country's leading space physicists. An early advocate of the use of artificial earth satellites in scientific research, he was the first Director of the United States Weather Bureau's National Weather Satellite Center, a position he held from 1962 to 1964, when he resigned to become Dean of the new School of Environmental and Planetary Sciences at the University of Miami. A graduate of Ohio State and Princeton, Dr. Singer has published a number of papers on the design and use of satellites, including their application to meteorology.

Thompson H. Mitchell has worked in the communications field for more than thirty-seven years. Now President of Radio Corporation of America Communications, Inc., he joined RCA in 1927, two years after graduation from the United States Naval Academy. Mr. Mitchell has played a leading role in the expansion and mechanization of the worldwide radiotelegraph system that has made RCA a leader in international communications. He is active and holds office in many professional associations in the electronics and communications fields.

James C. Hagerty has been in the communications field since he was a student at Columbia University. He was a political reporter for *The New York Times* until 1943, when he became press secretary to New York's Governor Thomas E. Dewey. He later joined the staff of General Dwight D. Eisenhower and was White House press secretary during the eight years President Eisenhower was in office. In 1961, Mr. Hagerty joined the American Broadcasting Company as Vice President in charge of news, special events, and public affairs. He was elected to his present position as Vice President in charge of corporate relations in 1963.

Richard B. France is a senior information officer in the Office of the Secretary of the Air Force. A specialist on Air Force research and development in missiles, space, and bioastronautics, he has worked closely with military and civilian scientists in these fields, both here and abroad. Mr. France is also a trained economist and expert in marketing and foreign trade, and his duties have taken him to twenty-eight countries in Europe, Asia, and the Middle East. He is the author of several articles which have appeared in national publications and scientific journals.

Dr. Abraham J. Heschel is a descendant of a long line of Hassidic scholars, the thirtieth generation of rabbis in his family. He was awarded his doctorate degree in philosophy by the University of Berlin and taught at the Frankfurt Lehrhaus as successor to the famed Jewish mystic, Martin Buber. In this country he served as Associate Professor of Philosophy and Rabbinics at the Hebrew Union College in Cincinnati, and is now Professor of Jewish Ethics and Mysticism at the Jewish Theological Seminary in New York. Dr. Heschel is the author of several books, including *Man's Quest for God, God in Search of Man, Man Is Not Alone,* and *The Prophets.*

Father Francis J. Heyden, S.J., theologian, explorer, teacher, and author, is also internationally recognized for his contributions to astronomy. An authority on the photo-electric method of observing solar eclipses, he has designed equipment, trained personnel, and selected observation sites for several Air Force-sponsored solar eclipse expeditions, which he has led in many parts of the world. Father Heyden has written many articles for professional and technical journals on astronomy and space research; he has also written essays on science and its moral and ethical implications. He is now the Director of the Georgetown College Observatory at Georgetown University.

The Rt. Rev. James A. Pike, J.S.D., S.T.D., is the Bishop of the Episcopalian Diocese of California and a brilliant theologian and teacher. Originally trained at the law schools of the University of Southern California and Yale, he left the practice of law to enter the church. He has served as chaplain at Vassar College and as head of the department of religion and chaplain at Columbia University. Bishop Pike is also a distinguished author and has written several books on the law, theology, and the political and social problems of our times, among them *Doing The Truth, Road Blocks to Faith,* and *Church, Politics and Society.*

John Paul Stapp, M.D., is Deputy Chief Scientist of Advanced Research at the School of Aerospace Medicine, Brooks Air Force Base, Texas. As a career officer in the United States Air Force medical corps with the rank of Colonel, he has made a specialty of seeking ways to provide for man's safety and survival in the air, in space and on the earth. It was for this purpose that he himself has served as the subject of rocket-sled experiments in which he was exposed to stresses of more than forty times normal gravity. Col. Stapp is a skilled pilot, a former president of the American Rocket Society, and winner of the Legion of Merit and numerous other awards.

SPACE: ITS IMPACT ON MAN AND SOCIETY

THE POLITICS OF THE SPACE AGE

Lyndon B. Johnson

The history of man is a history of triumphs, but a history of the governments he has devised to serve his needs is a history of failures.

Since the eighteenth century, man has substantially changed the nature and potential of his existence on earth with an outpouring of scientific inventions and discoveries. Over that same period, the casualty rate among man's political forms, systems, and values has been high. Of the eighteenth century's creations of political science, only the forms and values of the United States system of representative, constitutional government survived unchanged through the impact of the nineteenth century's Industrial Revolution to flourish and flower in the twentieth century.

This perspective is important. Any discussion of the "politics" of the Space Age is, necessarily, a discussion of politics in the classical sense, not the convention-and-campaign sense. If the

Space Age has not eliminated the smoke-filled room, it has at least opened a window and given air to fundamental and sometimes obscured values.

The fate of our free society—and the human values it upholds—is inalterably tied to what happens in outer space, as humankind's ultimate dimension. While the response of our technology is important, no less important is the response and role of our political institutions—both to the challenges of the present and to the opportunities of the future.

The vital role of politics in the Space Age has been evident since October 4, 1957, when the Soviet's Sputnik I was launched. The orbiting of that first unmanned earth satellite was a feat of science. But the worldwide impact and importance was essentially political. Sputnik I was proclaimed by the Soviets as validation of Communist preachments and prophecies about the superiority of their political system. Such conclusions have been proved premature and excessive by our own subsequent successes. But the fact remains that the Soviet "first" in space exploration resulted primarily, if not entirely, from misjudgments by our political leadership, not from deficiencies of our scientific community.

The price of underestimating space exploration has been much more tangible than a "loss of status" in world opinion. Whatever uncommitted nations and their leaders may have thought, the chief impact of Sputnik I was registered directly upon the Kremlin. Communism's leaders were convinced that the time of fulfillment for Marxian prophecy had come. They were emboldened to begin a period of belligerence and bluff toward the West which did not run its course until the abortive Cuban venture in 1962. Fiscally, we paid an enormous price—of which the Berlin buildup, for example, was only a small part—to compensate for the political misjudgments of the early 1950's.

The review of these events is as important for what it does not mean as for what it does mean. Most emphatically, it is not to be inferred that our free society should accept as its goals those set by the ambitions and undertakings of the totalitarians.

We would dishonor our trust to do so. The goal of free systems is human progress; the goal of the systems which compete against us is human enslavement. By definition, our work is never done. Human progress is a mountainside, not a mountaintop—political systems and politicians serving human progress may climb or fall, but never rest.

In our domestic American politics, this is the lesson learned from Sputnik I. After the sustained climb of our postwar "consumer goods" boom, some of our political leadership elected to rest, as though a summit had been attained. In that moment of illusion, we were passed by—in the technology of space.

Why did this occur? Self-examination is imperative. In retrospect, the answers seem to be these:

1. Despite two decades of intensive scientific advance, we had failed to establish adequate relationships between the scientific community and the political community—to the blame of each and the disservice of both. As politics disregarded economics in the 1920's with near disastrous results, so national politics in the 1950's resisted science with unfortunate consequences.

2. Inherent in the political miscalculation toward space exploration was the influence of the unfortunate anti-intellectualism of the early 1950's. American political leadership repeated the classic error of the profession through the ages: assuming that the summit of power is also the summit of wisdom. The open contempt with which the Sputnik I success was greeted by some in positions of political trust betrayed a degeneration of the respect for intellect which is indispensable to the governing of a free and open society.

3. Perhaps the greatest failure of all was devaluation of the first—and hardest—responsibility of elected representatives of the people in our system: the responsibility to lead. Elective office entails more than merely serving the comfort of the electorate. In the 1950's, higher values were attached politically to "pleasing" than to "pushing," and this uncharacteristic transposing of America's traditional values led to a lulling of national effort

and momentum toward the "new" and "unknown."

No purpose is served, at this late date, by excessive recrimination. Sputnik I jolted us out of the lethargy and bad habits into which we had drifted. Our political structure reacted responsibly. Science was accorded its rightful place as a resource and strength of integral concern to national policy-making. The blight of anti-intellectualism was erased from responsible political leadership circles—and, no less important, the scientific community lost much of its anti-politician bias which inhibited accord, understanding, and cooperation. Executive and legislative branches alike reassumed a responsibility for leading—for defining the higher goals and higher potentials of our society and system.

In the years since 1957, the American political system—and its politicians—have acquitted themselves well, as have our scientists. An orderly program of space exploration has been devised and financed without unbalancing the values of our free society. We have reached for the stars above us without forgetting the slums here among us. We have resisted the impulse toward totalitarian secrecy to conceal our failures, while also resisting the totalitarian monopolization of the knowledge acquired through our successes. These are significant achievements. Certainly it cannot be argued that the United States space program is guided by imitative response and reaction to the Soviet. On the contrary, the United States is clearly upholding the special and highest values of the revolution of freedom of which we are heirs and trustees.

Our most important achievement politically in the Space Age has been this retention of basic national character, purpose, and political values in our space program. It is reflected by a primary objective of our program, clearly stated at the outset, to gather knowledge for the benefit of all nations and to join with all nations in this endeavor. It is underscored by the fact that in the earliest days of the Space Age, the United States formally insisted that the exploration of space should be a joint adventure of nations with the express objective of fostering peaceful uses. This has been, and continues to be, our position both in word and deed.

It is a landmark achievement of our political system that the original National Aeronautics and Space Act authorized our government to cooperate with other nations in pursuing the legislation's basic objectives. During the past four years, we have cooperated with more than fifty other nations: sharing scientific information, training their scientists, and, in several instances, launching their space payloads with our rockets. Our communications and weather satellites have convincingly illustrated both the potential for peaceful uses of outer space and our unhesitating willingness to share those benefits. While this is only a first primitive stage, these developments underscore the fact that space exploration is the first human experience with the potential of joint participation by all nations on earth.

In my own judgment, this potential—which lies clearly in the political realm—of bringing nations together in joint productive common labors to share common benefits is one of the greatest potentials space affords—at least for the remainder of the twentieth century. Such common endeavors are basic to any real hope for peace. It is my conviction that space can be mankind's first real avenue to peace by giving all men and all nations a common self-interest and joint adventure.

This common self-interest is the hope of making a better world tomorrow for today's children. The recent United States and Soviet cooperative space effort in communication marks what may be an auspicious beginning for better East-West understanding through the advancement of knowledge. We would like to see this understanding broadened by joint space endeavors in meteorology, astronomy, and lunar and interplanetary exploration. For in the exploration of space, knowledge is both the key and the quest. When we reach for the moon and the stars, we reach for answers to the deepest secrets of the universe. How much more accessible these would be if we but strove upward together!

Since time began, change has been the mortal enemy of the

political systems man has devised to give order to his own existence, to safeguard his person and property, and to preserve and perpetuate the ideals of his culture. The Age of Space is to be one of the great eras of changing human experience; it is, perhaps, a mistake to restrict the concept of this new era by characterizing it as the age of "space"—it is, more accurately, an age of exploration and extension for all the fields of human knowledge and progress. In this age of increasing emphasis on science and technology, it is imperative that we not lose perspective on the challenge it presents to our political form, systems, and values.

America's special contribution to the world has been in the political realm. Our greatest success has been at devising a pragmatic political system capable of transforming spiritual ideals into realities for the lives of all our people. Progress among the segments of our total society is most useful when it serves to strengthen the total system. In the age now beginning, our scientific and technological progress can mean great strength for our political institutions, at home and abroad, but we must not displace democracy with an overriding technocracy.

We must commit our national resources to the support of achievements in space—as we have done. But that commitment must not be blind to balance and proportion. We must not proceed on the dangerously oversimplified logic that money alone is the root of all national successes—in space or other endeavors. That is a false premise, as is the premise that niggardliness is the root of all failures of our political system.

This new age is a time of growth for our national capabilities. Such growth requires the free system's politician—and the free system's voter as well—to grow also. We must all have greater breadth and perception, appreciating that the value of our new capabilities is the support and strength they give to our oldest values of freedom.

Domestically, the Space Age, like the Industrial Revolution, will mean dramatic upgrading of the promise and potential of the lives of all our people. If our political institutions are to serve the people effectively during this period, those institutions must function to improve the quality and availability of education, permit and foster growth and expansion of our economic system, and stimulate, by vigorous leadership and vision, aspiration toward higher national goals.

Internationally, our capabilities in this new age will permit us to exercise a new leadership of hope. The technical accomplishments of the Age of Space will be enemies of war. Improving international communications, for example, will serve to lessen the likelihood of misunderstandings leading to nuclear disaster. More importantly, the accomplishments of technology, in the not too distant future, will permit attacks upon human poverty, agricultural deprivation, illness, and illiteracy such as have never been possible in the past. We must exercise the inherent resiliency of our system to conceive meaningful applications of these new capabilities which will advance our goals of a world of peace, justice, and freedom.

Thus far, the Space Age has been characterized as a period of competition. It is important, however, that we keep in mind the fact that the competitiveness is between political systems, not between national scientific communities. In the world of science, the logical instinct is toward cooperation without regard to political boundaries. This impulse we must preserve. The real challenge of the Space Age is for the politician to tear down the walls between men which have been erected by his predecessors and contemporaries in the political field—rather than to raise its barriers higher into the free and peaceful vastness of space. If the potentials of the Space Age are fully realized, this period will someday be known—and blessed by all people on earth—as the Golden Age of Political Science.

EDUCATION FOR SPACE

James E. Webb

Science and technology today are more deeply involved than ever before in man's concepts of the universe in which he lives, and the opportunities and dangers which confront him as an individual and as a participant in organized society. He sees their mark on industry, business, and agriculture, and even on our routine day-to-day living. He finds them woven tightly into the social, economic and political structure of our nation and, indeed, of the world; and he perceives that they affect government decisions on national and international issues. As Dr. James B. Conant has said: "Whether we like it or not, we are all immersed in an age in which the products of scientific inquiries confront us at every turn. We may hate them, shudder at the thought of them, embrace them when they bring relief from pain or snatch from death a person we love, but the one thing no one can do is banish them."

The all-pervasive impact of science and technology demands,

now more than ever before, an educated society—educated both to understand and appreciate the workings of these forces and to direct them intelligently and morally. It does not mean that we must become a society of scientists. This is neither possible nor desirable. What it does mean is that our society must become scientifically and technically literate so that national programs and policies for space, defense, health, education, and welfare, as well as other endeavors in which science and technology are prominent, may receive the public understanding and support they merit.

Earlier in this volume, President Lyndon B. Johnson called attention to the political misjudgments that delayed our efforts in space. Certainly, one factor in these political misjudgments was the lag in the scientific and engineering literacy of our citizens. A public uneducated to the value of science and technology could hardly be expected to understand and support those members of the American scientific community who, more than a generation ago, were urging rocket and satellite development for the exploration of space. It took the burst of Soviet Sputnik I to compel us to re-examine our attitude toward science, to assess the capability of our engineers to design and build the space systems which were clearly required if we were not to lose our place at the table where international decisions are made, to question whether adequate emphasis was being placed on science and engineering in our educational systems and, indeed, to consider whether the generally accepted patterns of education did not require drastic modification.

Changes in education in the past fifty years have been limited largely to experiments in teaching methods and new teaching materials. It was assumed that these changes would release the creative talents of students and would encourage a wholesome self-expression which some educators believed was being stifled, if not crippled, by the more conventional methods and materials of instruction. But, in the opinion of many recognized authorities, among them Dr. Conant, this brought far too much emphasis on form and too little on substance. Certainly, with few exceptions, there is little evidence—particularly before 1957—that either the experimental schools or the conventional systems consciously attempted to develop an awareness of the substance of science and technology and a comprehension of their increasing role in national development.

Scientific literacy in general and scientific competence in particular are twin goals which a modern system of education must achieve. It is not fair nor even desirable to ask that the scientist assume sole responsibility for economic and social decisions in which, though scence may play its part, the collective judgment of all is indispensable. The scientist can provide us with the required theoretical and technical solutions for a particular space program, but its advisability may present difficult financial, political, and even social questions.

Some educators believe that education in science should begin rather early. Recently, at Columbia University's School of Engineering, for example, sixteen elementary school pupils eight to ten years of age participated in an experimental science course designed to guide children in discovering on their own the basic laws governing physical phenomena. By simple experiments, these young students learned and demonstrated the principles of gravity, electricity, and magnetism, and devised means to measure the combustion temperatures of various metals and a host of other materials. There is a wider growing belief that development of a scientifically aware citizenry should become as much an accepted part of the basic curriculum as reading, writing, and arithmetic. Some would even extend the initiation to the kindergarten.

Such early emphasis on science and its acceptance as a continuing program will undoubtedly result in an increase in the number of scientists, although it should be recognized that the truly gifted scientists are relatively few for many reasons—among them genetic makeup, motivation, and opportunity. Perhaps only a small percentage ultimately will find careers as professional scientists or engineers in colleges, universities, government, or industry. Others may become expert technicians and thus render yeoman service in the advancement of scientific knowledge. Still others may discover their natural aptitudes in administration, but all the better suited for their tasks because of their understanding and appreciation of scientific principles, techniques, and demands. A good many may find their self-realization in entirely different pursuits and specialties, but more fully qualified for civic responsibilities in an age of science.

Since Sputnik I, there has been some improvement in science education for the nonscientists, but it is not enough. Present and future generations will have to make their living in an economy supported by industries and professions in which technological advances are reducing the need for unskilled and semiskilled labor. The present paradox of rising unemployment while jobs are going begging will continue unless the new disciplines, skills and knowledge necessary to satisfy present employment requirements in the bulk of our industries are properly taught in our system of schools.

Expanded knowledge in science and medicine, as well as the high standard of living enjoyed by the large majority of the citizens in this country, have increased the life expectancy of the average American to over seventy years. Surveys show that except for the most highly trained, educated and skilled, people beyond the age of fifty are increasing the ranks of the jobless. Two-thirds of the people over sixty-five in our otherwise affluent society live in poverty, and this age group is increasing annually by 325,000.

For many of our youth the picture is equally disturbing. In this decade whose goal is man's exploration of the moon, a total of 26 million workers will be seeking employment. Half of this number will be under twenty-five years of age. In his special message to Congress on our nation's youth on February 14, 1963, the late President John F. Kennedy, on the basis of then current estimates, predicted that of this age group seven and one-half million will be high-school dropouts "unprepared for anything except the diminishing number of unskilled labor openings." In 1963, nearly one million young people between the ages of sixteen and twenty-one were unable to find jobs simply because they lacked the training—the education—to qualify them for the jobs available. A survey in the nation's capital showed that both equipment and curricula are outmoded in our vocational schools.

The educational lag in science and technology also exists in colleges and universities. According to practicing engineers, present engineering education in most professional schools is obsolete. These schools fail to provide their students with the basic scientific education necessary for the solution of problems the graduate will confront. Nor are most of these professional schools able to provide enough realistic exercises which can be applied in actual practice—especially for the large systems operations characteristic of most space work.

Advances in research and the massive fund of new knowledge also demand periodic re-education for professionals already at jobs. Dr. Edward Wenk, Executive Secretary of the Federal Council of Science and Technology, has pointed out that previously accepted levels of technical and scientific skill have become so obsolete that the "half-life," for example, of an engineer, once twenty-five years, is now something like five years. What this means is that education, at least in science and engineering, must become a continuing process, not only for professionals practicing their art or craft but for those who teach science and engineering. For the scientist and engineer and teacher, there is a definite challenge in living in a world where the rapid expansion and advancement of knowledge impose the need for further education and retraining. But it is frustrating and frightening for the growing number of adults displaced by technological advances who lack the literacy to learn the new skills they need in order to be re-employable.

A study reported on last year indicates that one-fifth of our economic growth is due to education. A recently published survey by the United States Census Bureau estimates a lifetime income difference of about $100,000 between those with no more than eight years of education and high-school graduates. It is evident that raising the quality as well as the quantity of our human resources by improved education is essential for our advancement on earth as well as in space.

The concept of education for all the people has been fundamental in this nation. It was inherent in the philosophy that sparked the American Revolution. In 1786, Thomas Jefferson, in a letter to George Wythe concerning matters of legislation under consideration, said, "I think by far the most important bill in our whole code is for the diffusion of knowledge among the people. No other sure foundation can be devised for the preservation of freedom and happiness. . . . Preach, my dear sir, a crusade against ignorance; establish and improve the law for educating the common people. Let our countrymen know . . . that the tax which will be paid for this purpose is not more than a thousandth part of what will be paid to kings, priests and nobles who will rise up among us if we leave the people in ignorance."

In this age of the atom and space, with knowledge accumulating at an explosive rate, education—the best education—for all our citizens is more urgently needed than ever before in our history. How much greater now is the hazard of ignorance than it was in the time of our Founding Fathers? And yet the fact is that present efforts in education at all levels are not meeting our national needs.

Improvement in education is primarily the collective responsibility of local and state agencies, but the stimulus and direction must come from the more than 2,100 colleges and universities in the nation, and to some extent from the Federal Government. There are forty-two Federal departments, agencies, and bureaus participating in educational programs to some degree. Major programs are conducted by nine agencies, of which the National Aeronautics and Space Administration (NASA) is one; but only two—the Office of Education and the National Science Foundation—have education as their primary concern.

Although education for the Space Age is not its primary goal, NASA does have substantial responsibility in this field, under the legislation pursuant to which the Agency was established in 1958.

NASA is specifically directed by the Space Act to take steps to increase the scientific and technical capability of the nation in fields needed for advances in space and to undertake "long-range studies of the potential benefits . . . and the problems involved in the utilization of aeronautical and space activities for peaceful and scientific purposes." Such studies have enabled NASA to assess the educational, social, and economic implications of its program. NASA's impact on education is in no small measure a result of its role as an employer of substantial magnitude as well as a customer for vast industrial goods and services that require unprecedented scientific and technical skills. NASA's interest in education is thus not peripheral or hortatory. It is essential not only for its immediate and specific tasks but also for the achievement of the wider national goals which hold so much promise for the future.

Education in science extends in many directions; and the educational programs, as sponsored or aided by NASA, vary with the need and purpose. Their common bond is the improvement and extension of the sciences, technologies, and allied disciplines. What NASA is doing is a small, but significant, part of our total federally supported effort in education. Federal assistance to educational institutions for facilities and equipment in fiscal year 1964 was less than a billion dollars, of which NASA's portion was $9,379,280. Military programs and loan programs under the National Defense Education Act provided support for undergraduate student aid totaling $150,000,000. Of Federal grants to graduate students, NASA's contribution was $19,811,604 out of a total of nearly half a billion; and the total number of students aided by all loans or grants was 470,000.

NASA's educational programs and services are generally aimed at college or university levels, but also include space-science materials for elementary and secondary schools to assist in updating classroom instruction and student participation; twelve spacemobiles—space-science classrooms on wheels equipped for lecture demonstrations which travel throughout the country and abroad; various educational publications and a speaker's bureau;

several motion pictures explaining space goals and activities; radio and television programs and more than 160 exhibits for display here at home and abroad. A pilot program to determine methods to present space science in adult education has been started. These materials and services are practical and visual means for supplementing general education in science by providing some understanding and appreciation of the goals and concepts of our space program.

Modern scientific research, in a sense, is something of a paradox. On the one hand, the vast extension of the frontiers of knowledge has required a greater degree of specialization by our scientists and engineers. On the other hand, much of our research, particularly in the space-related sciences, requires a broadening of horizons through the cooperation and support of many other scientific disciplines. Therefore, a good part of scientific research, to be most effective, must assume a group or interdisciplinary character.

Our universities are uniquely organized to marshal such wide diversity of professional talent, for they are the source for the trained manpower needed in all fields, including science and technology. Left to themselves, however, they are not likely to meet the ever-increasing and urgent demand for scientists, engineers, and technicians. Of the estimated 1,100 colleges and universities in the United States qualified to grant degrees in science and technology, only about 160 grant advanced degrees (Ph.D. or equivalent) in at least one field of science or engineering. Of these graduate schools, which constitute virtually the total national university research capability, only about twenty are recognized leaders—outstanding both in staff and facilities; and these have, in large measure, been the recipients of the greater part of Federal research funds.

There is no question but that this university research capability has contributed substantially to NASA's scientific and technological work. Most of the experiments carried aboard NASA and engineers within the university community. NASA also has satellites and deep space probes represent the work of scientists benefited from their advice and counsel in deciding on the kind of experiments and research to be undertaken. In recognition of the high quality of these contributions and the need to increase the number of cooperating institutions, in the fiscal year 1962 NASA established the Sustaining University Program. Under this program, training grants are awarded to universities for qualified pre-doctoral graduate students, and research and facilities grants to those institutions of proven ability conducting space research for NASA which require additional facilities. The over-all objectives are to increase the supply of scientists and engineers and to provide the conditions for improving research and training in universities and colleges.

Were NASA to make only the top twenty schools the recipients of Federal funds, it would not satisfy these objectives. However, NASA cannot put its scientific missions in jeopardy; nor can NASA undertake, as a primary program, the strengthening of weak universities. But it can and does make a conscious effort to seek out competence in other than the twenty leading universities. As a consequence, several other such schools are among the nearly thirty institutions which have been awarded facilities grants under the Sustaining University Program. NASA now has more than sixty small but important programs of research at as many institutions—all within the University Program. Thus, although NASA continues to use the facilities of the first-ranking institutions upon which much of our research effort still depends, it is broadening the national research base, particularly in the allocation of its training grants.

It was expected that the Sustaining University Program, through its training grants, would eventually assist in securing 1,000 Ph.D's annually and thus provide significant aid to meet the present and future manpower needs in space and other scientific endeavors. When the training program began, in April, 1962, ten grants were made to ten universities to support, over a three-year period, 100 graduate students working toward a Ph.D. In September, 1963, 786 students entered the program; and by September, 1964, an additional 1,071 will begin work on doctoral degrees in space-related fields under grants allocated by NASA to 131 colleges and universities in forty-seven states. The predoctoral training program is well within reach of its major objective to satisfy the nation's future needs for highly-trained scientists and engineers.

However, the Sustaining University Program was more broadly conceived, for it also contemplates and encourages the active collaboration of qualified specialists in the behavioral and biological sciences. A realistic program, especially in the space sciences, cannot afford to do otherwise. At the Space Science Summer Study conducted in 1962 at the State University of Iowa under the auspices of the National Academy of Sciences and sponsored by NASA, the agenda included "Some Social Implications of the Space Program." Among the topics considered were the effect of space-program expenditures upon national or regional development and on special sectors of the economy; practical applications of space science, the "spin-off" of new materials and new processes from the space program to other sectors of our economy, and the potential reception of such transfers, and the political and international aspects of space. If no definitive answers were reached, it is significant that the questions were asked and that there was general acceptance of the need to bring the continuing attention of the social scientists to bear upon the space program now. It is safe to say that, perhaps, only fifty years ago, a suggestion for such discussions at a scientific conference would have been regarded as a heretical trespass upon the sacred domain of science. Today the need is both understood and accepted.

This understanding and acceptance reflects a growing conviction by scientists as well as nonscientists that science and technology, whether for the conquest of space or of the atom, will be self-defeating if not enriched by research into their social and economic effects and possibilities. Such a wide and broad approach to scientific investigation is indispensable, and its importance must be taught and impressed upon those who will direct our scientific research in the future. The Sustaining University Program, though yet modest and in some respects experimental, intends to encourage the university to foster such collaborative efforts.

Education is the true road to emancipation for nations as well as men. Our nation, rich though it is in all resources, can no longer afford to waste its human resources by offering less than the best in education to all our citizens. The conquest of space requires an integrated human as well as scientific effort, a pooling of all our manpower and material assets, if it is to fulfill its bright promise for mankind's future.

Dr. Lee DuBridge of the California Institute of Technology, said in August, 1959, "One hundred years from now the new kind of knowledge attained in space research will surely have paid untold, unforeseen, and unexpected dividends. Already, the dawning of the space age had impelled Americans to seek to improve their schools. That alone may be worth the cost of all our space rockets." These remarks were made before our space program was as large as it is today, but I believe they are still true.

To warrant the expenditure of vast public funds and manpower, our scientific exploration of space, like Jacob's ladder, should, while extending heavenward, be firmly set upon the earth. Upon what firmer foundation can our endeavors be based than education?

LABOR IN THE AGE OF SPACE

Joseph A. Beirne

Are we not now, I sometimes wonder, living in a period of deceptive quiet and serenity—the first few seconds, relatively, of the Space Age? In our daily lives, few of us—except those most directly concerned with space planning or operations—are much aware of the Space Age problems that demand intelligent answers. Most of us merely watch the manned spaceshot on early-morning television before turning to the routine tasks of the day, thinking as we do so, "Spectacular, all right, but what's it got to do with me?"

That reaction, I suspect, is common throughout much of the business world and the labor movement. Most of us look at the space race as a kind of international and interplanetary Olympic Games. Will we get to the moon before the Russians? Will we land on some other star before voyagers from other planets visit us, or enslave or destroy us?

A few of us, with a better opportunity to peek through laboratory doors, have glimpsed the possibility of strange and awesome new weapons that may be dominant in the age of space. The laser light, some scientists tell us, is indeed the "death ray" of science fiction; they theorize that the laser, mounted on a manned space platform, will be the weapon to give its possessor nation ultimate and ubiquitous military control of the world through the threat of instant, pin-pointed destruction of any spot on earth.

Obviously, these are not the kinds of matters that are discussed or debated at Chamber of Commerce meetings or trade union conventions. The AFL-CIO convention of 1961 – four full years after the first Russian sputnik – did not deal with any resolution specifically identified as a Space Age problem. The closest it came was to denounce a proposal by Senator McClellan of Arkansas to ban strikes at missile sites; but both the motivation and the text seemed more a reaction to the Senator than to the Space Age. My own union, the Communications Workers of America, at its convention in 1963 was, of course, aware and proud of Telstar, the first working communications satellite. The delegates endorsed the union's position on a publicly regulated, privately owned American communications-in-space corporation. But both actions were essentially routine union business. First things first! Today's business today!

Yet labor and all the rest of American society must soon come to realize that the Space Age, with its new dimensions and its wondrous technology, will produce an impact of gigantic proportions. We cannot envision the exact details, any more than Columbus might have visualized the twentieth-century Manhattan skyline when he first sighted the New World. The space effort, with its demand for tremendously sophisticated equipment, calls for skills of innumerable variety. The magnitude of space requires that the challenging effort to penetrate it should also assume similar proportions – a vast coordinated mobilization of resources, technology, men, and money.

Already these problems have begun to move far back from the gantries and the launching pads, into the areas of the relationship between government and corporations. J. S. Dupre and W. E. Gustafson of Harvard University have pointed out that "The government has had to devise new standards in its contractual relationships with business firms. Essentially the government now assumes the financial risk involved in innovation. Free competition no longer characterizes the process of bidding for government contracts. While private firms have thus been freed from the restraints of the open market, they have acquired new public responsibilities. They are no longer merely suppliers to the government, but participants in the administration of public functions." This process, the authors note, has channeled public money into the corporate sector of the economy by administrative decision. They conclude: "Business, like government, must then become subject to non-economic checks to avoid abuses." If that reasonable assumption is correct, it is difficult to see how labor —by which I mean not only the trade union movement but, indeed, the entire work force—will avoid facing greater regulation in the decades ahead.

The leadership of organized labor thus has a responsibility to devote an increasing measure of its thinking to the adjustments in structure and attitude which the trade union movement may be called upon to make in the new space-inspired age of technology. The "old days" are indeed gone forever. A trade union movement, if it is to survive and to serve its membership and the community in the years ahead, must be prepared to change —perhaps change drastically—to stay attuned to the new times. Each passing month reinforces that belief. If American labor fails to adjust, it may indeed lose strength and relevancy, in which case the national community would be deprived of a constructive and important viewpoint.

To many Americans whose knowledge of labor-management relationships comes primarily from news story headlines, the question of labor's role in the Space Age is apt to produce the reflex response that the problem is essentially whether or not we will permit strikes in space industries. Our policy on strikes is important, of course, because strikes reflect a breakdown in communications and relationships between the two sides. But the problem goes further: to the composition and training of the labor force, the employment of the labor force, the outlook of the labor movement on social control of nongovernmental institutions, and the kind of society that the labor movement wants to see developed.

Contemporary America believes in the right to strike. When underpaid fish canners on the Eastern Shore of Maryland wage a strike to win a few pennies more than the minimum wage law provides, or employees of a service establishment in a city walk the picket lines, they are apt to win public support; and their efforts to hinder the employer's normal conduct of business are apt to be successful. But there is by now a well-established public assumption (sometimes incorrect) that wages and working conditions in big industries and corporations are "pretty good," and popular support for a strike may not be forthcoming. Equally important, strikers are having difficulty stopping production, at least in many industries involved in the new Space Age technology. Is it not symbolic of change that in the summer of 1963, for instance, Congress passed with little dissent, and the President quickly signed, a revolutionary bill to enforce compulsory arbitration on railway management and labor? The vocal reaction from business, the labor movement and the specific brotherhoods involved—all of which are "deeply committed" to oppose compulsory arbitration—was low-pitched and, indeed, lacking in rancor or spirit. Nobody appeared ready to challenge the assumption that the country did not want or could not tolerate a national railroad strike with its complex and unforeseeable effects on the whole economy.

In other cases, the new technology may blunt the desire to strike, or negate the effects of a strike if called. In the oil workers' strike at a Shell refinery in Texas, a rather small cadre of supervisory and engineering personnel outside the union's control was able to keep the automatic process running smoothly enough for months on end. My own union saw an almost identical situation in a strike against Southern Bell back in 1955; and from that experience we concluded that creative collective bargaining is a better approach. In a more recent strike by another union at an independent telephone company, the logic of our earlier judgment was apparently confirmed since, except for minor breakdowns and inconveniences, that telephone system maintained a reasonably high standard of operation.

These developments suggest that the strike weapon is losing universal effectiveness or automatic popularity with workers and public. The Space Age is apt to carry on this process. Certainly, workers in strategic centers of space work activity will be directly affected by this trend. Would society permit a strike to delay the departure of the daily rocket flight carrying supplies to a colony of men on the moon, or to interrupt the electronic communications with a military or nonmilitary manned space vehicle, or to delay the production of specialized foods consumed by men on space ships or space colonies? Rather, it would seem, the thinking that is concerned with these kinds of industrial relations problems must turn toward a variety of devices that will provide constructive and satisfying equivalents to the strike as a means of achieving constructive agreement. If the process is to be successful, the validity of the labor union must be accepted by business and government, and both union members and union leaders will have to realize that in specialized areas of the economy in the years of the Space Age, resort to the strike weapon may be socially unwise and organizationally suicidal. Yet if two nations locked for a generation in a far more furious struggle than any strike could reach agreement to ban nuclear tests that were poisoning the atmosphere around us, then it is reasonable to expect that labor and management, faced with equivalent dangers in the years ahead, ought to be able to find some way, jointly and voluntarily, to reach equitable understanding without the strike and the lockout. The price of failure will be very high.

Fortunately, there are hopeful signs. In a period that superficially seems much disturbed by strikes, the quiet successes get less attention: the steady improvements won for telephone employees by the Communications Workers; or the benefits to steel workers from profit-sharing at Kaiser and to automobile workers at American Motors; or the tremendous success of the government-inspired Missile Labor Board, which has corrected conditions that previously had caused a large amount of labor unrest in the missile industry. Granting flexibility and creativity, labor-management relations in the Space Age could prove to evolve more harmoniously than those of the troubled thirty years through which America has just been moving.

While we may learn to adjust the collective bargaining process to the demands of the Space Age, it may be infinitely more difficult to adjust the Space Age economy to produce full employment for the entire labor force at attractive income levels for every degree of skill and with a sufficient basis of personal security to permit and encourage the wholesome use of leisure time.

Our record to date does not suggest that we are close to that goal. The displacement of men by machines always has troubled wage earners and their unions, and today it is a paramount source of concern. George Meany, the President of the AFL-CIO, spoke the unanimous consensus of labor when he pointed out to the Congress that the problem of automation "is more important than balancing the budget." Calling attention to the fact that "we have now crowded into a few short years more technological change, more automation than we've probably had in our whole history," he has predicted gloomily that if the nation finds itself with a permanent corps of seven or eight million unemployed persons, the machine will have succeeded in destroying the "American way of life" and the enterprise system which serves as its foundation. Walter Reuther, in his capacity as Auto Workers president and chairman of the AFL-CIO Economic Policy Committee, has testified dozens of times before Congressional committees concerning labor's strongly held belief in government programs and industrial policies to encourage economic growth and achieve full employment.

Specific proposals have been advanced in recent years for

providing all our employables with jobs. Some are designed to share the work available: shorter work week, longer vacations, longer schooling, earlier retirement, specialized training and retraining. Other proposals envision legislative programs designed to accelerate the metabolism of the economic system as a whole. One conclusion inevitably emerges: at some point in the Space Age, we are going to be forced to mount a massive effort to develop a coordinated, effective economic program that will provide the employment opportunities the nation realistically requires. If we do not, the pressures will adversely affect the space program and our claim to leadership in world affairs and scientific exploration.

It is already becoming evident that the unskilled worker, the semi-skilled worker, and perhaps even many of today's skilled workers face a bleak future of unemployment or marginal employment. Our economy offers less and less opportunity to the man or woman of few skills. Thus, the American Negro population, so many of whom have experienced inferior education or discrimination in hiring or promotion, has been correctly interpreting the trends. A strong factor in the Negro protest movement has been the realization that lack of skill, or opportunity to acquire skill, can at this moment in our history lead him to an empty life in the wasteland outside the high fence that surrounds the affluent society.

Many workers—regardless of their race, color, creed, or sex—are discovering as they reach the mid-point of their work careers that jobs which they considered "good for life" are disappearing on short notice, under the impact of automation and the new technologies. In a few rather well-publicized cases, collective bargaining has won partial security for them; they are retained on the payroll until retirement in one capacity or another, but no new people are hired. Retraining programs have had only moderate success. In the communications industry, we have achieved a fortunate balance between the growth of the system and the productivity of the employees. The number of jobs in the Bell System has not notably changed for several years, although there is no guarantee of what the future may produce.

Serious as the job problem is today, it is bound to become even more difficult. Until the government, under the aegis of Secretaries of Labor Arthur Goldberg and Willard Wirtz, launched the public campaign against school dropouts, the public seemed apathetic to the fact that no less than 40 per cent of our young people were leaving school before high-school graduation—at the precise moment when the skill requirements of industry were undergoing a vast mark-up. At a time when we are becoming more conscious about the need to conserve our natural resources, we have been altogether too blasé about the fact that 51 per cent of the boys and girls in the top two-fifths of their high-school graduating classes do not go on to college. Their failure to do so may well be depriving us of trained brain power that will be crucially needed twenty years from now, or even sooner.

Just as the hard facts of Space Age technology are causing us to re-examine our notion of how many hours should constitute a typical work-week, so the same hard facts should be convincing evidence of the desirability of a radical adjustment in our educational program. We should, for example, be moving to establish a compulsory, free, public junior-college system for every student capable of absorbing higher learning. Legislative action should be taken to increase the "normal" range of public school attendance from the present twelve years to fourteen—perhaps even fifteen. Final graduation from the public school system—with possible exceptions at both ends of the aptitude scale—should be from junior college. And while I do not speak as an expert on the development of curriculum, it would appear essential that in these extra years of compulsory schooling which I strongly recommend, emphasis should be given to history and economics on one side and the language and disciplines of the Space Age on the other, namely, mathematics and the sciences.

The effects of the trend in the work force from blue collar to white collar and white coat, as the nature of the industry changes and raises the demands on the employee for better educational preparation, are certain to be felt by the trade union organization. Just as the language of the early American union seems quaint and archaic to our ears, with its reflection of Midwest populist and European socialist philosophies, so the trade union language of today—indeed the whole "image" of the union—will be forced to evolve to meet the needs and tastes of the next generation of workers.

The union, which was formed by immigrants or sons of immigrants who had little access to formal schooling and who had to fight hard on the picket line for recognition and progress, will have to look and sound a lot different to appeal to the young technician straight out of junior college or technical insttiute. Without flexibility in its approach to the member, to the industry, to the community and the philosophy of our society, no union will be able to survive very long in the Space Age years ahead.

That process of change will develop, because most unions reflect the will and viewpoint of their members. Changes in the educational, social, and economic composition of the membership inevitably affect the election process. Many of the early leaders of the CIO unions, for instance, were unable to make the transition from the role of agitator and strike leader of the 1930's to the position of negotiator and administrator as the unions gained security and maturity. Those who could not change fell by the wayside. Those who had greater clarity of vision and freedom from encumbering ideological harnesses led their unions to continued achievements. The Space Age is apt to be a similar period of trial and choice.

The race into space is obviously not a thing apart from our national society. The space program directly affects jobs, collective bargaining, civil rights, education, housing, the health of people, the health of democratic institutions, and the entire economy. Some people see the space effort as a prowler who steals the money that otherwise would be earmarked for health, education, and other social projects and are understandably worried that the glamour of the space effort may undermine these needed down-to-earth humanitarian projects. The layman has no reliable guideposts to lead him to wise decisions concerning the proper balance among allocations of governmental funds.

Irving Ferman of the International Latex Company, writing in *The Washington Post* in August 1963, reminded us that just as the great powers in other eras maintained their world leadership by domination of the seas, "our nation will maintain its position as a world leader for peace and freedom in large part by its dominance of space. The choice to commit so much of our resources to the Apollo Project is an honorable and moral choice for which all our citizens can feel the deepest sense of national pride."

American experience in the twentieth century has long since demonstrated that our freedom can be protected only by strength. Our best chance for peace in space and peace on earth is to develop steadily a scientific space program that will command the respect of all other peoples. We need both guns and butter—both space and social welfare programs. We can afford them both, and we cannot afford to sacrifice either to the other. That implicit assumption, it seems to me, underlies the lack of great discussion about space in the American labor movement. We instinctively support the space effort. Only now are we beginning to realize the vast economic and social implications of the space program for the multitude of us who have no present intention of riding a rocket to the moon or to Mars.

Far from being considered a drain on the American economy, the space program should properly be regarded as an incentive both for economic growth and social progress. The investment is already starting to produce returns. Dr. Louis Dunn of Space Technologies Laboratory, in an analysis of ten years of space from 1959 to 1969, reported in testimony before the Congress that "a careful analysis of the relative costs of satellites and of more conventional communications systems" showed substantial economies through the use of satellites even as early as 1959. Dr. Francis W. Reichelderfer, the former chief of the Weather Bureau, has estimated to the Congress that the value of more accurate weather predictions, made possible through space satellites, would be in the billions of dollars and would result in great savings in life and property. Dr. S. Fred Singer's account of weather in space in this volume gives an exciting insight of the fulfillment and promise of our weather satellite program.

But the economic and social benefits go beyond these narrow "profits" of space development, important as they may be. The space effort is a collective effort of the entire society, the entire economy. It is, in fact, a moral substitute for military struggle. Its aim is life and knowledge, not death and destruction. The impact of spending for the Space Age may be broadly felt throughout the entire economy—an economic improvement that spiritually is planes above the level of "war prosperity" or the meager benefits of a depression-born made-work program.

Will we in labor and, indeed, all of us in America have the wisdom and the fortitude to move steadily toward the rewards that the Space Age holds for us? Who can read the crystal ball? Yet, the American labor movement, I believe, has the intelligence and instinct to meet the challenge. Today's unions in this country thus have a compelling responsibility to start thinking about the nature of the changes they must comprehend in the Space Age. Failure to keep pace may mean the end of the union. Without some form of legitimate and independent trade-unionism, repression or unreality are apt to move in to fill the organizational void. Success in keeping abreast of new times will help assure that no vacuum occurs in the job of representing the practical short-run needs and the longer-range obligations of the working people; if the unions keep pace, it will be a positive gain for America in the Space Age.

SPACE—THE GOVERNMENT AND THE ECONOMY

Stanley H. Ruttenberg

Space exploration is inextricably tied to the national economy. Measured against an economic yardstick, our space effort holds great promise for the future, but falls short of success at the present time. For every million dollars spent today on missiles and space, the resulting direct employment is considerably less than for a million dollars spent a decade ago on the production of tanks, ordnance equipment, and piston-driven aircraft. Expenditures for urban renewal, mass transportation, and housing and educational facilities have considerably greater effect on employment than space spending.

This does not mean we should curtail our spending on missiles and space. Such expenditures, quite apart from our program objectives, are amply justified by the great technological and scientific advances they stimulate. But we must not overlook the need for increasing the level of private and government investment in other sectors of the economy to meet the social needs which thus far, paradoxically, have been intensified by the space effort. For technology and automation, both in the space field and in the general civilian economy, are displacing workers. Fewer job opportunities are available today than during the 1950's.

Every study of future employment patterns indicates abundant demand for the skilled and specially trained, and a greater decline in demand for the unskilled. It is therefore necessary to train and locate those entering today's labor market and to retrain and relocate those displaced through technical advance. The real problem is not to identify the needs but rather to determine how they are to be met, particularly in the space and defense industries.

These industries grow by government demand and necessity. When certain items are needed for our national advancement, government demand is met both by industry and labor in the areas best equipped for producing the necessary items. When these demands are satisfied, there is a curtailment in both industry and employment. Communities may acquire vested interests, as well as industry and labor. Government, therefore, has a special obligation to help in the necessary transition when a particular government program is at an end. This requires a partnership between industry, labor, and government to plan ahead—many years ahead; but the responsibility for making the partnership work is, in the main, that of government.

More money is being spent by industry for training and retraining than is being spent by government. But industry's motive is profit for its stockholders, and its training programs are necessarily limited in extent. They are not meant to be the answer to the total problem.

It is not the employer's responsibility to train displaced workers unless he plans to employ them. No employer should be required or forced to train any individuals for jobs that will not exist in his plant but may exist in other industries and in other geographical areas. Nor should the unions or the workers themselves be obligated to provide resources for retraining. But, though the government has a special responsibility, government alone cannot do the whole job. It will need the assistance of industry and labor—the one to provide employment opportunities and to assess present and future manpower needs, the other to encourage and stimulate incentive for training and retraining.

An approach to a more comprehensive solution to the paradox of growing unemployment is the Manpower Development and Training Act, passed by Congress in 1962. Under this Act, the Department of Labor is to undertake the training of over 400,000 workers within the next three years, for which task the modest sum of about $500 million has been appropriated.

The Department of Labor's recent experience with the Manpower Development and Training Program shows the magnitude and the wider ramifications of the task ahead. In community after community, for example, to organize a class of twenty for training as chefs, it was necessary to interview and test more than two hundred unemployed individuals to find twenty who were literate enough to read the instructions for recipes. Almost three hundred to four hundred persons had to be interviewed to fill a class for twenty or thirty for training in auto mechanics. The lack has been not so much in the required intelligence as in the rudiments of reading and arithmetic.

As our technology advances, the demand for unskilled workers declines, while the need for semi-skilled, skilled, and professional people—technicians, engineers, and scientists—is growing. An effective training program for our future labor force therefore requires better and longer schooling for our youth and a vocational training system designed for current and future needs. For the presently displaced labor force and for those whose jobs are threatened by technical advances, the problem is more immediate and pressing. A minimum adult education program must accompany or supplement the planned training programs for chefs, for male nurse's aides, for automobile mechanics, for mechanics to service typewriters or computers, and other needed skills.

In the long range, the advancing technology in space and defense work offers us as yet unmet opportunities. This technology is a threat only if we fail to lift the level of performance and direction. We must take the time and talent necessary to think not only about the civilian uses but the social implications of the technology we are bringing to the fore in our space and missile programs. These broad perspectives were recognized, in part, by Congress when it created the National Aeronautics and Space Administration, which is directed to explore the social implications of space development and translate our successes in space in a way that will extend and promote economic growth and meet our needs in health, education, and welfare. We can afford to advance in both directions. To do one without the other would only diminish our success.

This is a fact that business as well as labor must understand and that the individual taxpayer must accept. It is the failure of such understanding and acceptance that accounts for the growing pressures to reduce the level of government spending while pressing for the advancement and expansion in space research and exploration. Putting these two pressures together poses an anomaly that is a matter of grave concern to all. There are some who advocate that we increase our expenditures in space and allied efforts and neglect or ignore the social and economic effects and dislocations of these efforts.

Business representatives and other elements of our society who oppose federal funds for social needs do so allegedly because they fear such support extends the power of the federal government. To them almost any such extension is either unsound or suspect. There is no need to debate here the merits of this position. It is both short-sighted and oddly one-sided so far as it concerns space and defense industries, which are largely dependent upon federal contracts for their economic vitality. Large capital outlays by these industries require a supporting "infrastructure" of allied services and facilities. If these industries are to function at all, there must be available an adequately trained labor force, and today this means largely skilled labor properly trained. There must also be available adequate roads and transportation facilities, housing, health, educational, and recreational facilities. It is doubtful that our space and defense industries can or even desire to meet these indispensable social costs. To a very large extent these social costs must be met either by the other contracting party, the federal government, or by local government with federal support. No other choice seems feasible if the federal government's responsibility for full employment under the Employment Act of 1946 is to be attained.

Where government sponsorship of space and defense facilities is direct and immediate, its responsibility is correspondingly increased. More than mere building of an installation is involved; an entire community is being created. For example, the Mississippi Delta valley, which till now has been mostly swampland, is being redeveloped to become a new industrial center for the testing of the large space boosters needed for manned lunar and interplanetary exploration. It was selected because it provides the waterway (the transport roads) to carry the giant rockets from the gulf coasts of Texas, Mississippi, and Louisiana to the Atlantic coast of Florida where Cape Kennedy is located and where the giant rockets someday will be launched carrying three men in an Apollo spacecraft to the moon. Adequate transportation for the rockets eventually will require the building of the equivalent of the Panama Canal.

The construction of these space facilities is already having a tremendous industrial impact on this southern region; and this is only the beginning. As the facilities become operational, there is no question that much of the space-development industries in Southern California, northern Washington, and mid-Kansas will be curtailed.

But as the concentration of our space program is shifted to the Mississippi Delta area, we cannot ignore the problems that will be created in these other industrial areas. It is the government's responsibility to see to it that other industries are attracted into these western states. Around these large industries communities have grown, families have settled, and schools have been built. There must be a massive effort to maintain the continuity of these communities. Whatever it may cost will be less of a burden than that caused by the dislocation of whole communities and the destruction of individual as well as community security. This is a challenge for programs like the Area Redevelopment Administration established during the administration of the late President Kennedy.

In such federally supported communities, the government is not only a dispenser of public funds but also the employer of a vast labor force. It is therefore the business of government to be concerned with the rights of labor and the protection of those rights; for these also are essential for the maintenance of a healthy free-enterprise system. These include collective bargaining and the right to strike. Whether we can develop techniques for avoiding strikes in our missile and space industries because of their effect on national security is another question. This may, perhaps, be accomplished by developing new collective bargaining techniques.

The Missile-Space Board has been encouragingly successful in solving and resolving many labor disputes. Only one or two end up in strike situations for the many hundreds that develop from time to time but are resolved without strikes. There is every indication that industries and unions are seeking ways to find voluntary methods to solve their disputes in order to avoid interference with the national security needs. But even if this should not be possible, the right to strike should be protected.

The reason for labor unions today is not just to increase economic benefits for the worker. The union also is the instrumentality that protects each man's dignity on the job and provides an opportunity for him to voice his grievance and get a just hearing. This is as important as rates of pay; for how one is treated on the job and whether he can fight injustice and discrimination is vital to industrial democracy.

The business of government in space then is more than to secure its program objectives. It must do this and at the same time, working with industry and labor, see to it that human and social interests are considered and protected. More federal spending may be required to achieve this larger purpose; but it will be well worth the investment if it assures that technology becomes the servant rather than the master of men.

THE AEROSPACE INDUSTRY: RESPONSE AND RESPONSIBILITY TO OUR NATIONAL GOALS

Martin Goland

The scientific and technological revolution which began only twenty years ago was sharply accelerated by the launching of Sputnik I in 1957, and the establishment of space exploration as a major undertaking by our government. A direct outgrowth of our space effort, the aerospace industry was a response to an unprecedented national demand, and in a few short years it has grown into a significant and vital force on the American industrial scene. The impact of the aerospace industry may best be measured by its direct contribution to our space program and its influence on American industry as a whole.

It is estimated that in 1964 the aerospace industry will contribute in excess of $24 billion to our gross national product. More than $18 billion is allocated for missile and space activities within the Department of Defense: $6.5 billion for research, development, test, and evaluation; $7.6 billion for missile procurement; and $4.2 billion for aircraft procurement. Expenditures of the National Aeronautics and Space Administration during this same period are estimated at $4.4 billion. The projections for the fiscal year 1965 indicate that NASA expenditures will rise to about $4.9 billion. The Department of Defense budgets will decrease to around $16.6 billion. Incidentally, neither the DOD or NASA include the expenditures for the procurement of commercial aircraft which, in 1964, are expected to exceed $2 billion.

Aerospace spending in 1964 will thus account for around 3.9 percent of the $610 billion gross national product, and for about 4.2 percent of total manufacturing employment. For purposes of comparisons and to illustrate the magnitude of this infant industry, it may be noted that in 1963 the contribution of both the gas and electric utility industries to the gross national product amounted to some $20.5 billion. Total expenditures by the American public in the same year for food and alcoholic beverages is estimated at $86.7 billion, roughly only 3.5 times our budget for aerospace production.

A nation which subscribes to free enterprise is committed to the premise that all of its economic activities must be profitable. The missile and space industry, supported by public funds, must therefore share the burden of proven profitability. A 1962 survey of fifty-one aerospace companies shows a total invested capital of $3.76 billion, and a return on investment of 9.6 percent. To the financial analyst, an annual return of 9.6 percent is an indication of industrial health and vigor; and from this point of view the industry's performance is quite creditable. Some may suggest that the proper measure of an adequate return is the risk of investment and that in the aerospace industry the financial burden of product failures is borne by the government. Nevertheless, there are those who point to the low profit margin of the industry when measured by net income as a percentage of sales, which in 1962 was a modest 2.4 percent. It is argued that this low profit margin is government-enforced, and that it stifles the growth of individual companies. It is further suggested that the measure of return should take into account that aerospace goals are a challenge to the seemingly impossible and that each new step forward adds new dimensions of excellence to our industrial capabilities, which provide the key to future industrial growth.

The overriding factor, however, appears to be that the aerospace industry is unique. Some two-thirds of the industry is totally concerned with the military defense of our nation. The remaining portion constitutes the so-called "peacetime" or civilian space program. Both segments of the industry have but one customer—the United States government. In the normal pattern of private enterprise, the entrepreneur risks the future for the privilege of profit, and it is profit which affords a tangible measure of success. In the aerospace industry, the ultimate risk is not borne by the industry itself, but by its customer, the government it serves. Success or failure is measured by the extent to which government policies are implemented and national aims achieved.

Such an industry is public industry, even if served by private corporations. The criteria which govern the performance of the aerospace industry are not those of the free marketplace, but those of the conference room, the negotiating table and the ballot box. Serving a single governmental master, it is a captive industry, given subsidy on the one hand and left unprotected on the other from sudden shifts in the political winds. Justification for the aerospace industry thus must be viewed in terms of its contributions to national goals. With respect to national defense, which accounts for two-thirds of the total program, argument is obviated by the present state of international affairs. Most of us would agree that the continued survival of the Free World depends on our maintenance of an invincible military position.

It is rather with respect to the role and purpose of the civilian space program that basic controversy continues most acutely. Even to those with unbounded enthusiasm for our nonmilitary space goals, the enormity of this national effort must give cause for reflection. Never before in the history of science and technology has so massive and systematic an effort been launched. Equally unique historically are the complex circumstances which brought it into being.

The development of the Vanguard scientific satellite was

proceeding at a modest rate when the political shock born of Sputnik I suddenly propelled us into the large-scale space race. For the most part this immediate response was dictated by the injured pride of a nation which had come to think of itself as pre-eminent in all material achievements. As the program gained momentum, this understandable but superficial reaction was succeeded by a surging tide of wonder and a universal dream and promise of exploring the heavens. But such excitement of high adventure does not persist too long, and is bound to be followed by more sobering reflection. Increasingly, the public's attitude toward the space program is changing from that of the enthusiastic patron to that of the prudent and calculating investor. As underwriter and dispenser of subsidies, the public is demanding an accounting in the light of which our aerospace goals and expenditures may be appraised. This demand and appraisal are eminently just.

It should be stressed that the question is not one of absolute choice between competing and irreconcilable demands upon public funds. It is really one of relative priority and emphasis, since the substantial contributions of the aerospace industry to our national economy must be acknowledged if intelligent inquiry is to be fruitful and constructive. The major difficulty in the process of appraisal stems from exaggerated claims and counterclaims.

Perhaps the most convincing argument made by enthusiasts who support the investment in our space program is that the massive collection of new skills and advanced knowledge generated in the course of space activities will also be directly applicable to other branches of industry. In the long list of disciplinary fields commonly accepted as branches of the physical and biological sciences, scarcely one is omitted from the range of space concerns. Speaking before a recent NASA conference, an executive of a large aerospace corporation pointed out that in 1943 his company employed professionals in some seven fields, but in 1963 more than 175 different scientific and engineering disciplines were represented on the company payroll. With interests so broad and versatile, extending from new techniques in production engineering to the life sciences, it would seem that space-generated knowledge holds potential profit for every segment of American industry.

There are other benefits from the space program, and not the least of these is to education. Our space effort is exerting a positive influence at all levels, from the elementary school to the graduate college. Space activities are encouraging the training of increased numbers of highly trained professionals, and these will benefit the entire spectrum of American industry. The space industry characteristically needs the most advanced skills in the laboratory and the office and on the production line. Industry and the government clearly recognize that they must share in the responsibility of insuring an adequate flow of trained personnel to satisfy their needs. This, incidentally, is a departure from the classical thesis of free enterprise that the worker must bear the sole responsibility for making his services economically attractive.

There are some who view these developments with mixed feelings, for in our technological advance they see a threat to job opportunity. Automated machines, they fear, will replace human operators, and increasing plant productivity will reduce the number of available jobs. This pessimistic outlook ignores the experience of the aerospace industry and its creative output. The aerospace industry is a vast complex of machines and men. Its researches and process of manufacture demand equipment, components, and raw material which are obtained from other sectors of the economy. It is also a substantial employer of scientists, engineers, technicians, and others from our labor force. In 1963 its sales amounted to over three percent of total goods and services that comprise our gross national product. Four percent of our manufacturing work force, about one million three hundred thousand persons, is today engaged in tasks created by aerospace sciences and engineering within the past two decades.

There is no doubt, however, that the pace of industrial change in the space age will result in readjustments within our labor force. Continuous training and education, and the retraining of those displaced from older areas of technology in order that they can serve the new, will be the order of the day. "Technical obsolescence" of the professional man is a subject of vital current concern, and the issues raised by automation are already being discussed at the national level among labor, industry, and government leaders. The space industry has helped to call attention to these problems, and through its own educational experiments it is helping to point the way to future solutions. But the aerospace industry alone cannot solve the problem of training and retraining. Only through a concentrated joint effort by labor, industry, and government can we approach the necessary solutions.

The civilian space program is also contributing significantly to the nation's area development. The major centers of NASA research and development are being dispersed across the geographic face of the nation. Where these centers are established in regional areas in which industry has been slow to develop, the result is a sharp encouragement of the over-all local industrial activity. For example, NASA's Marshall Space Flight Center at Huntsville, Alabama, was established in a region whose productivity was limited to produce—notably water cress—and some cotton. Ten years ago the city of Huntsville itself was a small agricultural community with an area of three square miles. Today the area, whose major economic strength is built around the space facilities, extends to fifty-six square miles. This growth is reflected in a significant increase in professional offices (doctors, lawyers, accountants), shops, businesses, churches, and schools. School facilities have increased at the rate of a classroom per week for the past seven years. The University of Alabama Extension at Huntsville is fast becoming an important center for research and learning in engineering and in the physical and mathematical sciences.

The space complex at Huntsville is complemented by the Michoud Operations near New Orleans and the Mississippi Test Station at Gainesville in Hancock County, Mississippi. The Michoud Operation, fifteen miles from New Orleans, is located in a former shipyard which used to manufacture cargo planes during World War II and tank engines during the Korean conflict. It was completely idle from July 1953 until September 1961, when NASA selected it for space production. The once empty yard is now a large manufacturing center with more than 9,000 employees, working either for NASA contractors or for the government. The growth is reflected in new suburban communities, new housing and businesses, more schools.

In Hancock County, Mississippi, about thirty-five miles from New Orleans, on 13,500 acres of land that once was largely cow pasture, NASA's Mississippi Test Operations is under construction. When completed, it will employ 2,500 permanent persons, many of them with professional or technical skills. Plans are already under way by community leaders to provide for the needs of newcomers in housing, education, and other essentials.

NASA Launch Operations Center at Cape Kennedy has made its mark on the surrounding areas, which have developed from quiet empty stretches of sandy beach to varied and booming communities. The region reflects the excitement and vigor of a flourishing frontier.

The Manned Spacecraft Center at Clear Lake, Texas, in addition to its direct dollar contribution to the Houston economy, has had the far more important result of sparking a broad advance in the industrial interests of the community, until now overdependent on petroleum and natural gas. Already, 124 national companies have opened new offices in Houston to maintain liaison with the Center; manufacturing and servicing facilities will inevitably follow in many cases.

The Center presently employs 3,500 persons, nearly half being engineers and scientists representing skills which up to now had existed in the region only in a few small university departments. Some 75 percent of these professionals and their families came to Houston from other areas, and the impact of their arrival is being felt in virtually every channel of community activities. Throughout the Southwest, university curricula are being strengthened and enlarged to meet the challenge of new interests. Professional society activity is on the increase; primary and secondary school systems are being revitalized.

The most powerful argument in support of the economic benefits of the civilian space program is that the "spin-off" of advanced technology from space to other areas of industry is a rich national resource. Space-gained knowledge, according to this view, will open the way to new products, more jobs, and higher productivity. In discussions of the industrial values of new technology, a dramatic (but by now repetitious) tactic is to list the multitude of companies, both large and small, whose principal products were laboratory curiosities one or two decades ago.

The early efforts to define the value of space science and engineering to industry at large were based on the optimistic hope of establishing a direct connection between commercial benefits and their space-inspired origins. Perhaps the most spectacular illustrations are satellites for weather forecasting and communications. For the development of a commercial communications satellite system a new corporation has been established with half ownership by the United States government. The unorthodox investment pattern developed for exploitation of communications satellites and the public clamor for participation illustrates once again that space industries do not fit the normal pattern of free enterprise.

Space technology is contributing enormously to the design and development of high-speed, large-scale digital computers. Computers for use in spacecraft and missiles demand the ultimate in compactness and reliability, and these same characteristics are valuable for commercial application. The net result has been a steady flow of knowledge, with a minimum time lag between the development of new space-computer techniques and their subsequent appearance in a new generation of commercial products.

Microminiaturized electronic components are today off-the-shelf items as a consequence of a similarly rapid "spin-off" process. Developed primarily to meet space and missile requirements, the technology is now serving a variety of high-quality civilian uses. One of the more important and growing uses for "space electronics" techniques is in the field of medicine, where a revolutionary class of diagnostic and clinical tools are emerging. Bioengineering has more than humanitarian appeal for the scores of industrial companies now actively engaged in its early exploitation; it also holds the promise of substantial and profitable markets.

Production techniques developed by the aerospace industry also are finding their way into commercial applications. Among these are improved techniques for welding and soldering and explosive forming (the use of an explosive-generated shock wave in water to form large and heavy metal parts by expansion against an outer die). Synthetic lubricants and gas bearings, which originally were designed for space use, are equally valuable in industrial machinery and instruments. In the home, plastics and ceramics developed for rockets and missiles are used in new cookware and other products for the housewife.

The real significance of "spin-off," however, cannot be evaluated by the simple expedient of listing new products and processes which have space origins. To justify a research and development program costing many billions of dollars in this manner, it would be necessary at least that the path of transfer follow a straight line, readily traceable from space source to industrial use. As a practical matter, the path is usually erratic, circuitous and time-consuming, without a definite starting place or final destination.

The most direct influence will be felt in the so-called "glamor" segments of industry. These are the new, science-based industries founded on the most recent technology, with large research and development staffs. Receptive to striking out in new product directions, these are the industrial groups which have the best capabilities for speedily transferring technology from space to commercial markets. Computer and electronics companies have been rather successful in making use of "spin-off." Aerospace companies, on the other hand, lacking an adequate management orientation toward commercial marketing, have as a general rule been unsuccessful in their attempts to diversify from military and space products to general commercial products. Long used to thinking of product cost as a secondary factor to performance and reliability, they have been unable to meet the rigors of competitive pricing and to make the necessary adjustments.

For the great bulk of American industry the real meaning of "spin-off" can only be found in the general advancement of the technological state-of-the-art to which it contributes. Perhaps the first comprehensive study of this subject was conducted by the Denver Research Institute under NASA sponsorship. Its report, published last year, includes that "the more subtle forms of technological transfer have had, and will continue to have, the greatest impact—not the direct type of product transfer which is most often publicized."

In scores of commercially important areas, such as energy conversion, semiconductors, new materials development, instrumentation and techniques of management and production control, the space contribution forms but a part of the steady onward flow of progress. The fuel cell, for example, is a primary battery useful in space applications because it is the most efficient known means for converting the chemical energy of fuels into electricity. First conceived at the beginning of the nineteenth century, it offers enormous potential for earthbound applications which range from superefficient central electric-generating stations to economical individual electric-energy sources for the home, the factory, and as motive-power for automobiles and other vehicles. Its basic principles have long been understood, but the technology needed to produce practical fuel cells for these applications is still well in the future. In the current effort to develop fuel cells for the admittedly highly specialized needs of spacecraft, the over-all technology is being advanced and the future nonspace application brought nearer.

The Denver Research Institute Report emphasizes that "a time lag exists between the development of technology for primary missile/space use and its commercial application. Large expenditures on missile/space programs have been made only in recent years and there has not been sufficient time for many transfers to take place. It is highly probable, therefore, that most of the transfer is still to occur." In the years required for the slower-moving segments of industry to absorb these benefits, the space origins of the contributions will no doubt be obscured.

If the true meaning of "spin-off" can only be found in an advancing state-of-the-art, our expenditures for space science and engineering should be viewed simply as part of an accelerated national research and development effort. National expenditures for research and development in the United States currently total approximately $18.5 billion per year, representing slightly more than three percent of the gross national product. Included in this three percent is the cost of the civilian space program, presently around $5 billion per year, and thus amounting to around 0.85 percent of the gross national product.

Few would quarrel with this allocation for research and development. If anything, it is probably too low. The manufacturing segment of American industry finds it necessary to invest over four percent of its gross sales in research and development to protect its technological future. One may view the civilian space program, therefore, as advancing our national research and development effort to a proper level.

The critics of the civilian space program may argue that scientific personnel and funds now directed to space can be used with far greater effectiveness if directed toward solving the pressing problems in the world about us. The list of useful projects which can be advanced with an annual research budget of $5 billion, or even a fraction of this amount, is almost endless. Perhaps of greatest interest to the industrialist is the proposal that national funds be employed to upgrade the technology of segments of American industry labeled as "backward" and presently in the throes of economic regression.

These arguments are in a sense plausible, but not as definitive answers. First there are 65,000 highly trained professionals in the civilian space program. Though large in absolute terms, it is relatively small when compared with the 1.7 million who make up our total complement of scientists and engineers. Secondly, space exploration has become a national goal of the American people. It is supported because of its total impact—political, scientific, educational, and industrial—which the nation finds valuable.

It should also be recalled that technology knows no industrial boundaries. Students of the research process have long since given up their attempts to relate the planning and direction of a research program with its ultimate effects throughout the economy. In too many instances, it is found that the benefits are unrelated to the motivations which brought them into being. Science and technology are fluid commodities, and they will flow from space-oriented research and development laboratories to other sectors of industry with the same ease and facility that has always characterized the flow of new knowledge and the interchange of novel ideas. The material advance made possible can serve an enlightened nation and when blended with wisdom and humanity, new doors to a brighter future will be opened.

SPACE AND DISARMAMENT

William C. Foster

In two decades the surge of modern science and technology has vastly increased man's power and means of destruction. In this headlong rush man has been propelled to the fringe of space, and as the exploration of this "new world" advances, the question of whether the East-West arms race will be extended into orbit inevitably arises. A resolution adopted by acclamation by the United Nations General Assembly on October 17, 1963, provides a benchmark of intent in this connection. In that resolution the United States and the Soviet Union joined other members of the United Nations in calling on all states to refrain from placing weapons of mass destruction in orbit and thereby in affirming the view that such an extension of the arms race can and should be avoided.

The United States support of the resolution reflected previously existing national policy respecting the use of outer space. It was also in accord with our broader desire to bring the arms race to a halt, to eliminate the threat of war, and to establish the condition for a secure and peaceful world. It represents part of our nation's effort to devise arms control and disarmament measures which can provide a sounder basis for the protection of our security interests than is offered by the further multiplication of the techniques of destruction.

As long as our national security continues to rest in large measure on the strength of deterrent force, new scientific and technical advances will have to be examined to determine the opportunities they may present and the hazards they may pose. Outer space is no exception. The opening of outer space was in part a by-product of the ballistic missile. Moreover, both the United States and the Soviet Union acquired the basic ingredients of orbital nuclear delivery systems as the result of early advances in the technology and techniques of spaceflight. So, it is understandable that there should have been much speculation as to whether "he who controls space" could control the earth.

But at the present juncture of the arms race, there is no possibility that either side could gain decisive military advantage by placing weapons of mass destruction in orbit. Nuclear weapons orbiting in outer space would not render obsolete the existing types of deterrent capabilities. Indeed, it is generally recognized that the "bombardment satellite" would be more costly but less effective than the ballistic missile. This is likely to remain the case for some time to come. Thus there is no reason to suppose that deterrence has to be "in kind" to be effective. The deterrent effect of the ballistic missile is not limited to other ballistic missiles, and there is no basis for believing that a force of "bombardment satellites" would be deterred only by a similar capability.

Although such a force would doubtless have considerable psychological impact, its military effects would, on the whole, be limited. The nation possessing such a force would still have to face the prospect of coping with an opponent's missile forces if it were to initiate a nuclear attack from outer space. And, from what we see in the picture today, it is not likely that the offensive striking power of one side would render the other unable to retaliate, or that its defensive capabilities would present an impenetrable barrier. Accordingly, the "bombardment satellite" would represent a marginal capability, and the military incentive for either side to acquire it will be low for the foreseeable future.

Of perhaps greater significance is the interest both sides should have in avoiding the hazards that would accompany the introduction of the "bombardment satellite." If forces of nuclear armaments have grown for want of a better alternative, then at least this growth has been attended by major continuing efforts to reduce the risk of war by accident. This requires that nuclear weapons and vehicles for delivering them must be reliable and subject to control under all conditions. A spaceborne capability would present substantial difficulties in this regard.

Although the United States and the Soviet Union have achieved remarkable advances in spaceflight, each has encountered problems with the reliability of some spacecraft. The failure of spacecraft engaged in scientific research and exploration constitutes an acceptable risk. But unreliability in a "bombardment satellite" would be quite a different matter. Is it not clear that the standards would have to be much higher if failure or accident might endanger the lives of millions of people on earth? Would the risk be acceptable if the price of failure or unreliability was the initiation of war by accident?

Neat theoretical solutions for ensuring reliability and control of a spaceborne nuclear force may be designed on paper, but the time when such a force could be deployed without risk of accident seems far distant. Despite its awesome potential, the ballistic missile can be more readily kept under control and the risk of accidental war that much diminished.

Such strategic and practical considerations were carefully weighed by the United States. It concluded, as former Deputy Secretary of Defense Roswell Gilpatric stated in September 1962, that the placing in orbit of weapons of mass destruction would not represent a national military strategy for either side. These same considerations also suggested that the interests of both the United States and the Soviet Union would be served by avoiding the drain on resources and the hazards to security that a nuclear arms race in outer space would entail. The Soviet Union expressed its recognition of this community of interest in September 1963, and, subsequently, it joined the United States and the other participants in the Eighteen-Nation Committee on Disarmament in sponsoring the so-called "no bombs in orbit" resolution.

Ambassador Stevenson, commenting on United States support of the resolution, called attention to the fact that it is not possible to foresee today all events which may at a future time occur in the newly emerging field of space technology and in the exploration and use of outer space. For this reason, if events as yet unforeseen should suggest the need for a further look at this matter, the United States would bring such events to the attention of the United Nations. Also for this reason, the United States does not plan to let down its technological guard. However, the resolution does remove some of the uncertainty which has in the past precipitated new cycles of the arms race. Together with the agreement reached in 1963 on a limited test ban treaty which prohibits nuclear weapons test explosions in the atmosphere and under water as well as in outer space, the prospect of avoiding an arms race in space has become more promising.

The nuclear test ban and the no-bombs-in-orbit resolution are steps that have been taken to contain the arms race. But another aspect of that race is the formidable problem of reducing the high levels of armaments already existing in national arsenals.

The dual role of the rocket as a booster for ballistic missiles and as a space-vehicle booster presents one of those ambiguities arising from the basically neutral character of technology. If it were possible to reach agreement on reductions of armaments, including reduction of ballistic missiles, would continued testing of advanced space boosters offer a route to improved ballistic missile capabilities? Could the production of space boosters serve to offset reductions in armaments? Would an effective program of disarmament be incompatible with an effective program for the exploration and use of outer space? It is questions such as these that require the disarmament planner to exercise all the skill at his command.

Missile and space booster technology are integrally related. There is little question that testing boosters for use in space activities could, at least in some respects, provide data applicable to military missile systems. Agreements should be made to exclude some possibilities of this character. But controls over the testing of boosters—short of the unacceptable solution of eliminating testing completely—should not be expected to cope with all such possibilities. Accordingly, a primary concern of safeguards accompanying a disarmament agreement would be to bar the application of advanced techniques to operational military systems. It is against the emergence of such operational capabilities and not against technological advance as such that disarmament arrangements would be directed.

This leads, in part, to consideration of production aspects of the problem. Certainly, there should be assurance that ballistic missiles and other armaments were not being produced in contravention of an agreement. But it would be important as well to preclude the accumulation of excessive stockpiles of space boosters which might provide a basis for rearmament. To curtail space-booster production and impose an arbitrary limitation on space activity would not provide a satisfactory answer. A better approach would seem to be to provide for an appropriate balance between production rates and planned and actual rates of space launchings. Production could then be conducted at the levels warranted by the current level of space activity.

An additional safeguard would be to destroy missile launching facilities as missile forces were reduced. Re-establishing such facilities if an operational missile capability were sought would require extensive efforts and an extended period of time. These factors would act as a brake on any effort to rearm through the use of available space boosters or through the conversion of production lines left open in support of space programs.

The complex interrelationships between missile and space programs are not likely to be resolved through a single approach. However, with arrangements addressed to the curtailment of the qualitative arms race and to the reduction of existing armaments, such approaches should have a mutually reinforcing effect. In combination, these approaches should reduce the problem to manageable proportions.

Through the research and planning effort of the Arms Control and Disarmament Agency (ACDA), the United States is proceeding to clarify these matters further and to develop the detailed techniques required. These efforts are not intended to hinder scientific and technical advances but rather to ensure that such advances are channeled into peaceful applications. The objective is to secure the necessary assurance without imposing unnecessary burdens on the continuing exploration and use of outer space.

Among these continuing programs would be space activities in such fields as communications, meteorology, navigation, mapping, and geodesy. Some space activities of this character may, in fact, contribute to the achievement of arms control and disarmament objectives. Acquisition and transmission of data by space vehicles may offer a means of clarifying situations where uncertainty, doubt, or misinterpretation might increase the risk of the outbreak of war through accident or miscalculation. Limiting this risk is a prime objective of arms control during a period when levels of armaments are high. Space systems also may well be capable of supplementing ground verification procedures in support of any future disarmament agreement.

Clearly, the relationship between arms control and disarmament, on the one hand, and outer space, on the other, is by no means one-sided. Although space activities pose certain problems from the standpoint of arms control and disarmament, they may also make a special contribution to arms control and disarmament objectives. While it is important to take steps to rule out the possibility of placing weapons of mass destruction in orbit, it is equally important to proceed with the development of other space capabilities, such as those related to observations from space, which can play a constructive role prior to and during disarmament.

In sorting out the over-all complexities of this twofold relationship (entailing restraint in certain space activities and encouragement of others), ACDA has had the continuing cooperation of the Department of Defense and of the National Aeronautics and Space Administration. Expert assistance and advice have been made available by both. ACDA, through contracts, has also had available the experience of scientific and industrial concerns which have been active in the ballistic missile and outer space fields. Such concerns have taken an active and constructive interest in ACDA's research program and planning effort.

Activities of NASA warrant a special mention here, for they have a bearing on arms control and disarmament progress in another and broader context. It is evident that mistrust and suspicion among nations act as barriers to progress toward arms control and disarmament. Cooperation in scientific and technical endeavors may be one of the most effective ways of improving understanding and creating the basis for mutual trust. For this reason, the steps toward cooperation in space research which have been taken by NASA and the Soviet Academy of Sciences, and the more broadly international steps being encouraged by the United Nations Outer Space Committee in the technical and legal aspects of outer space, assume a special importance. If these steps can be successfully pursued and gradually extended, they will increase the range of opportunities for the two nations to work together and with others. Hopefully, this kind of cooperation will help remove a number of misconceptions that may now exist on both sides.

While the arms race continues, there may, of course, be some limits to the extent and character of space cooperation. But as space technology continues to advance, as the population of space vehicles in orbit increases, and as new types of space vehicles such as large space stations in sustained orbit are introduced, increased cooperation may prove to be a productive as well as a convincing means of obtaining (and providing) assurance that the technological surprises that may lie ahead in outer space are being turned to constructive ends. As a general proposition, it appears reasonable to expect that the greater the degree of cooperation, the narrower will be the range of problems requiring attention from the standpoint of arms control and disarmament.

No useful purpose would be served by encouraging a belief that progress toward disarmament will be simple or accomplished without difficulty. But early success in this task is important not only to release the world from the shadow of self-destruction but also to free resources which now must of necessity be devoted to defense. Studies by the ACDA and other interested agencies are already under way to ensure an effective conversion of our industrial, human, and material resources. And research by some of our major defense industries has demonstrated that disarmament is *economically* feasible, a fact also borne out by a review of our industrial development and the economic adjustments demanded by the changing circumstances in this century.

The ability of the American economy to adapt itself to meet national needs has been proved in two great wars as well as by our response to the Soviet challenge in the conquest of space. Our industrial capability provided the weapons that enabled us to win two wars—neither of our making. It has also developed the machines and instruments to advance our peaceful goals in space for benefits which all mankind may enjoy. Our pursuit of these goals provides a path to the economic adjustment that disarmament will require; for it offers an important outlet for the persons, property, and production that will be released from defense activities. Inspection and enforcement systems, whether by individual nations or by an international disarmament organization such as the United States has proposed in its draft treaty, would involve not only land and sea, but air and space as well. Surveillance by satellites would be an essential part of such systems.

Disarmament will also enable us to divert the manpower and resources now confined to maintaining our military supremacy to beneficial uses of space materials and techniques. Communications satellites, weather control, improved means of air transportation, and the economic exploitation of the moon and planets are promises that may be fulfilled in a world without arms. Defense industries could build the machinery and spacecraft for the mining and transportation of certain metals expected to be found in relatively pure form on the lunar surface. This might prove a profitable alternative to the costly refining and separation processes which we now must use to obtain these metals in pure form. The manpower now used to build armies as well as armaments may—by the next century—be employed building colonies on the moon and other bodies in our solar system.

There are, of course, the great social needs which still remain unsatisfied. When war is outlawed among men and nations, we can better gather our forces and resources for an all-out crusade against poverty, disease, and that greatest enemy of all mankind—ignorance. Our Space Age technologies, now necessarily devoted to defense, can provide the weapons for this grand assault. Scientists have said that if the men and machines now otherwise employed in gathering and analyzing data for defense could be concentrated on basic medical research, the fundamental causes of all diseases could, perhaps, be discovered in a generation.

The challenge to master outer space and the ancient enemies of poverty, ignorance and disease that threaten us on earth is great; but the challenge to bring the arms race under control, which is really the challenge for man to master himself, is even greater.

DOES THE MILITARY HAVE A ROLE IN SPACE?

General Bernard A. Schriever

History has shown that every medium which affords military possibilities has been exploited for military purposes. This has been the case for land, sea, and the atmosphere; and in view of the achievements and the ambitions of our opponents in world affairs, there is little reason to believe that space will be an exception.

In a book on military strategy published in 1962, Marshal Sokolovskey of the Soviet Union wrote: "Soviet military strategy takes into account . . . the use of outer space and aerospace vehicles." Since the Soviet manned orbital flights in 1961, Soviet leaders have frequently boasted that they could easily use their space technology to build "global rockets" or place 100-megaton bombs in orbit.

The Soviet objective is world domination, and to their way of thinking, space weapons—no matter how costly or implausible they may appear at this time—might be regarded as useful means toward this end. Such weapons—or threats to make use of such weapons—are ideally suited to the Soviet tactics of blackmail and attempted intimidation. This grim possibility serves as a reminder to us all that space is not remote.

Our nation has demonstrated that we cannot tolerate the presence of offensive weapons 90 miles from our shores. Yet Soviet Sputniks and Vostoks have traversed the United States at altitudes ranging from approximately 100 to 150 miles. For the first time in our history, the vast expanse overhead has been penetrated by vehicles we can neither identify nor intercept.

In order to guard against the possibility of such a threat to our national security, it is necessary to investigate the military potential of space systems and to determine the magnitude and effectiveness of potential threats. If an effective threat were possible, we could not afford to neglect it. As long as it may be possible for our opponents to find ways of using space for aggressive purposes, we cannot afford to neglect the development of the military space technology that includes means of identifying, inspecting, and achieving rendezvous with unknown objects in orbit.

It is clear that military spacepower may not only be important for the over-all defense of the United States but also will supply by-products to science and technology that will support the scientific experiments carried on by the National Aeronautics and Space Administration. These experiments, utilizing both manned and unmanned vehicles, are designed to expand human knowledge of phenomena in the atmosphere and beyond for the benefit of all mankind.

The question has been raised as to whether the conduct of both military and civilian space efforts does not result in unnecessary duplication of effort and increased cost. Why cannot the same systems serve both civilian and military requirements? Why, for example, have two communications systems? The answer is that military systems because of the nature of their operations, demand more urgent development and higher performance standards than their civilian or scientific equivalents.

Similarly, civilian systems may have their own unique requirements which are not met by military development programs. A military communications system, for example, may need to carry relatively few messages, but it must offer the highest reliability and security and must be able to operate under extremely adverse conditions. A civilian communications system is not required to meet such exacting reliability and security standards, but to be economically feasible it must be able to carry a much higher volume of messages than a military system.

The unique requirements for military operations in space include, among other things, the capability for fast reaction and repeated missions, the ability to conduct missions at times and places dictated by national defense objectives, the capability to inspect uncooperative objects in space, the ability to survive and operate in a hostile environment, and, in general, the same kind of flexibility the Air Force now possesses in operations within the atmosphere.

These requirements are far more demanding than those imposed by a program with solely scientific objectives. A scientific space probe, for example, can wait to be launched until conditions are ideal. A space vehicle with a defense mission cannot wait until conditions are perfect. Moreover, many of NASA's experiments are designed for one-time special missions—the exploration of Mars or Venus, for example. Military satellites, on the other hand, must be designed for routine repetitive operations around the clock.

Our military space efforts have a twofold objective: to strengthen the over-all defense of the United States and to protect the specific interests of the nation in space. In the years just ahead, we will concentrate on the region of space near the earth. At present there are no programs for military space developments beyond a distance of about 22,300 miles from the earth's surface—the altitude of the synchronous satellite which will stay fixed in position above one point on the earth's equator. We cannot rule out the possibility that there may be military requirements beyond this distance, but for the time being we are primarily concerned that a hostile power does not dominate near-space.

Space systems can enhance the defense of the United States by increasing in many ways the military capabilities of our terrestrial forces. For example, they can provide early warning of a missile attack and perhaps ultimately they may even afford an active defense against ballistic missiles. Surveillance weather satellites can report meteorological conditions, and communications systems can improve the response and provide the command control of our forces.

We also need space systems that can protect the specific interests of the nation in space—including its right to conduct peaceful experiments and exploration—by permitting us to know what is happening in space, providing ways to determine whether a threat is present, and affording methods to counter potential threats in the space region.

One of our primary military goals in space now and in the immediate future is to gain familiarity with the military characteristics of space through experience in space operations.

Among the specific requirements of military space technology are the development of reliable and relatively economical launch vehicles suitable for use in routine military space operations; development of devices for both sustained low-thrust and quick-reaction high-thrust propulsion in space; achievement of capability for rendezvous, docking, and transfer; development of in-space power supply systems; development of secure and reliable communications systems which can function during periods of natural or man-made interference; acquisition of the knowledge, equipment, and techniques needed to transport and support man in space and to permit him to function effectively there; and the development of techniques for precise re-entry of manned vehicles into the atmosphere in a manner that will permit landings within the largest possible geographical areas.

Some of this technology will be acquired in cooperative programs with NASA and other agencies. The X-15 research rocket plane, which is a joint Air Force-Navy-NASA program, has yielded much useful data on aerodynamic heating, operational and control problems, hypersonic aerodynamics and structures, problems of exit and entry, weightlessness and physiological reaction of pilots to these conditions. The successful sub-orbital and orbital flights in the NASA Mercury program have corroborated our expectation of manned space flight. Likewise, the Department of Defense should gain valuable knowledge and experience from joint participation with NASA in the Gemini program, which will explore the possibilities of prolonged manned space flight and the rendezvous of space vehicles.

But valuable as this experience may be, NASA programs by themselves will not build a military capability. This is not their purpose, nor should it be their purpose. A military capability can be created only by a military organization which possesses a combination of technical knowledge and operational experience with suitable military equipment. Both NASA and the Department of Defense have valid and distinctive roles in the national space program. Their efforts are complementary, not competitive; their programs are cooperative, not conflicting.

The close cooperation between the Air Force and NASA is striking proof that preparation for national defense in space is not inconsistent with the national policy that space be used for peaceful purposes. In fact, it is an implementation of that policy because it provides the means for insuring that the policy is carried out.

Therefore, military development will remain a necessary and vital part of the national space program. These military space efforts, in addition to helping provide for national security, also promise to accelerate progress toward attaining all of the national objectives in space. This is a logical prediction in view of the fact that the Space Age itself, to an extremely large degree, is a result of military technology. Much of today's space technology and hardware have derived directly or indirectly from research conducted for military purposes. Just as the phenomenal growth of aviation in this century has been spurred by the exigencies of war, so has the rapid advance of space technology been stimulated and fostered by the requirements of national defense.

In order to see these requirements in perspective, it is helpful to recall briefly the growth of the aerospace threat to our national security. The experience of World War I indicated that the atmosphere could be used as a ready means of access to enemy targets, and accordingly airpower became a major factor in the planning and execution of military strategy. The progress of airpower became so rapid that even before World War II, the scientists in Nazi Germany were at work on a weapon that would supersede the airplane. Their efforts resulted in the V-2, a rocket-propelled ballistic missile.

Although the V-2 did not attain altitudes much in excess of 100 miles, it clearly foreshadowed more powerful missiles that could travel through space during a major portion of their trajectory. Immediately after World War II the Soviet Union, exploiting the talent of German scientists and utilizing its own considerable experience in rocket propulsion, began an intensive program aimed at developing an intercontinental ballistic missile. This program served as the foundation for the subsequent growth of Soviet space technology.

The Soviets have continued to make progress in the exploration of space, and as long as they continue to be controlled by an aggressive and militant ideology, their achievements in space constitute a source of potential danger to peace and freedom. The military aspects of the United States space program represent measures taken to respond adequately to the military and technological dimensions of this challenge.

Our experience in various aspects of missile and space technology extends over more than forty years. As early as 1918, Major H. H. Arnold, who became Commanding General of the Army Air Forces during World War II, participated in experiments with a primitive guided missile nicknamed "The Bug." Twenty-six years later, as Chief of Staff, General Arnold called together a group of the nation's top scientists—headed by the "Father of Aerodynamics," the late Dr. Theodore von Karman—and asked them to think about the needs of the Air Force twenty years in the future. Among the bold new ideas conceived by this scientific group was a serious study of "world-circling space ships"—presented to Air Force leaders in 1946—which concluded that earth satellite systems were feasible.

Research into the biological and medical problems of space flight is also built on a foundation created in the closing months of World War I. In January, 1918, the first U.S. military aeromedical laboratory was established to study the problems of exposure to low temperatures, wind-blast, vibration, acceleration, and the lack of oxygen at high altitudes.

This laboratory, now called the Air Force School of Aerospace Medicine, has long been recognized as a pioneer in the study of the medical problems of high-altitude flight. In 1949, the School established its Department of Space Medicine, the first serious research group at any scientific institution to be devoted to the practical requirements for manned space flight. In 1951, the School sponsored a symposium on "The Physics and Medicine of the Upper Atmosphere"—the first major international meeting on space medicine.

The early interest in space shown by aeromedical researchers was very natural; many of the physiological and psychological aspects of space flight are closely identical with those encountered in atmospheric flight at high altitudes. Biologically speaking, the space environment begins relatively near to the earth's surface. At a few thousand feet the effects of diminished atmospheric pressure begin to become evident, and by the time a pilot has gone much above 60,000 feet, it does not much matter, physiologically, how much farther he goes. At about 63,000 feet or above, the atmospheric pressure is so low that without the protection of a pressure suit or pressurized cabin a pilot's blood would literally boil. At altitudes much above 73,000 feet, manned aerospace vehicles must carry their own atmosphere in sealed cabins, since the outside air becomes too thin to be compressed efficiently.

During the early 1950's the Air Force conducted experiments in a number of significant areas of space research. In 1951, weightlessness experiments were conducted with mice and monkeys carried to an altitude of 35 to 40 miles by Aerobee rockets. The same year more than thirty weightlessness flights were achieved in fighter-type aircraft at Muroc, California, and Wright-Patterson Air Force Base, Ohio. These flights followed the "Keplerian trajectory," in which the earth's gravitational pull is counterbalanced by centrifugal force, producing a brief period of weightlessness. This technique has since been used extensively for research in weightlessness.

In 1952, the Aeromedical Laboratory at Wright-Patterson Air Force Base began centrifuge experiments to see whether humans could stand the acceleration forces of earth escape velocity. Other studies of acceleration and deceleration forces were conducted in an extensive series of rocket sled runs on the tracks at Edwards Air Force Base and Holloman Air Force Base.

In 1954, the Air Force gave the highest research and development priority to its ballistic missile development program—the most comprehensive military development program ever undertaken by the United States. This program not only developed the missiles that are a vital part of the nation's deterrent forces, but also created much of the technology, facilities, and hardware that have made our national space program possible. These have been utilized extensively by the Department of Defense in its space efforts and by the National Aeronautics and Space Administration, the Atomic Energy Commission, the Weather Bureau, and the other agencies which are involved in space activities. Most of the nation's space shots have been launched by a military booster such as Atlas, Thor, or Agena, or by a modified military booster such as Delta (modified Thor).

The national space program demands the cooperative efforts of all the agencies engaged in space efforts. For many years the Air Force worked closely with the National Advisory Committee on Aeronautics to advance our nation's capabilities in aviation. When the Committee was replaced by the National Aeronautics and Space Administration in 1958, the cooperation continued.

Air Force support to NASA includes procurement of Agena,

Atlas, and Thor boosters and contractor technical and launch services in support of such projects as the Mariner interplanetary space probe, the Ranger lunar probe, the orbiting observatories, and the Echo passive communications satellite; procurement of Titan II boosters for the Gemini two-man orbital spacecraft program; conduct of wind tunnel tests for the Apollo three-man spacecraft, and other programs; conduct of range operations, maintenance and logistic support services for the Manned Lunar Landing Program; central procurement of propellants for NASA research and development tests and space launches; procurement for NASA research and development efforts in bioastronautics, aerospace medicine, radiation studies, and human tolerance studies; and maintenance of facilities at its missile ranges.

One indication of the magnitude of military assistance to NASA is the fact that the Air Force, Army and Navy together provided 24 ships, 126 aircraft, and 18,000 people in direct support of the first Mercury manned orbital flight. For Major L. Gordon Cooper's successful 22-orbit flight in May, 1963, this support was increased to 29 ships, 171 aircraft, and more than 18,000 people. In addition, nearly 200 military officers are assigned to duty with NASA, and military facilities throughout the country provided extensive support to many phases of the nation's space program.

NASA space efforts also have benefited from early Air Force development work in a million-pound thrust engine and pioneer studies in such areas as lunar landing, long-range communications, guidance, high energy rocket fuels, and re-entry vehicles.

In addition, military developments in space have yielded benefits to the safety and physical health of the nation. Cryogenics—the study of the effects of extremely cold temperatures; bionics—the use of living models as a key to the functioning of electronic and mechanical systems; solid-state physics; and molecular electronics are only a few of the new specialized scientific disciplines fostered by military space research. Tiros, the research weather satellite, is a significant example of the civilian application of military research and development.

The advance of space technology in itself will not automatically produce a better world. But spacepower can form the protective umbrella under which we work for better ways to solve the problems of mankind. As the late President Kennedy pointed out in dedicating the Manned Spacecraft Center in Houston, Texas, "Space science, like nuclear science and all technology, has no conscience of its own. Whether it will become a force for good or ill depends on man, and only if the United States occupies a position of pre-eminence can we help decide whether this new ocean will be a sea of peace or a new terrifying theatre of war."

United States military capabilities in space must insure that no nation achieves a position in space which threatens the security of the United States. It provides an essential guarantee that space research will be carried on in an atmosphere of freedom and that the scientific knowledge which results can be used to benefit all mankind. This is the promise and challenge posed to the military by the Space Age.

THE LAW IN OUTER SPACE

Nicholas deB. Katzenbach

Many fascinating questions about the law of outer space have been posed by legal scholars, political scientists, and others. To the thoughtful layman, however, the question that would probably seem most important for mankind's survival and progress is whether the protection of a suitable system of space law really exists today. In other words, until there are policemen in space or other means of law enforcement, can we correctly think of space law as real law that will be observed—something more than just theories, opinions, and proposals?

The simple answer to this question is yes, we do have space law today, law which is well beyond being purely theoretical. In fact, there are two kinds of space law, although both are necessarily still in their infancy. The first kind is national space law—law affecting space activities which each nation can impose through its power over its own territory, or over its own citizens.

National law can regulate aspects of space activities that have a home base or other connection inside the nation. A nation may promulgate its own law and apply it to activities affecting space engaged in by anyone—citizens, aliens, or even men from Mars —if such activities are conducted within the territory of the nation. As an example of national space law based on such territorial jurisdiction, the United States can and does regulate the sending of radio signals by anyone from within this country to communicate with men or satellites in space, as well as signals transmitted from this country which might interfere with space communications. By the same token, national law may regulate the launching and landing of space vehicles in the territory of the nation, and may control the disposition of space debris or fragments of space vehicles which fall on its territory. The United States has taken the view that such fragments should be examined in order to determine ownership and then returned to the nation of origin. This policy was followed recently when a chunk of a Russian sputnik landed in Wisconsin.

National law may also apply to some activities conducted far out in space, if engaged in by persons owing allegiance to that nation. This is similar to the present-day application of some of our federal laws to United States citizens anywhere in the world —for instance, the federal law on treason, which applies to our citizens to the ends of the earth. There seems no doubt that this law would also apply to treasonable conduct in space by United States citizens. As another example, there is little doubt that Congress could apply our income tax, if it chose, to the income earned by United States citizens while on the moon, regardless of the future status of the moon. The same might be done in certain other fields—perhaps immigration and nationality—to clarify, for instance, the legal status of a child born on a space platform. Solutions to such a question, when needed, might be guided by the situation of children born on the high seas under present law, where the nationality of the parents governs.

National space law, whether based on nationality or on territorial jurisdiction, may be either new or old: either specially created rules to regulate activities in or affecting space, or long-established rules that are merely applied to space activities because they seem to fit. It may also be a blend. An example of national space law which blends both old and new elements is the Communications Satellite Act of 1962. This law is designed to create an effective organization for United States leadership in establishing a global system of communications by satellite, for the purpose of better and cheaper global telephone, television, message and data communications. The opportunity for human betterment is great, because better, less expensive communication, like better, less expensive transportation, is a handmaiden of the progress of civilization. At present, long-wave radio is often unreliable over great distances, cables are expensive for their limited capacity, and micro-wave circuits cannot go around the curvature of the earth without relay points, which the satellites would provide *par excellence.*

The new law is based on a proposal submitted to Congress by the Kennedy Administration, after careful government study by the FCC, NASA, the State Department, the Justice Department, and the Space Council. The law weaves together several established legal concepts, such as the private business corporation, government regulation of utility companies (which include communications common carriers such as telephone and telegraph companies), and positive safeguards to preserve and encourage competition among the suppliers and the customers of the new system. The essentially novel element is a three-way linkage in which the communications industry, the federal government, and the public at large are harnessed together positively to make the program succeed. The chief mechanism for harnessing these forces is a unique private corporation with six communications industry directors, six directors chosen by the investing public, and three appointed by the President of the United States. Supplementary machinery for promoting and guiding the program is provided in the form of additional government responsibilities to see that the corporation functions as contemplated. The whole organzation will be capable of fitting into an international system for satellite communications, through international arrangements which are yet to be developed.

Satisfactory international arrangements are of course essential to the communications satellite system, since no one nation is in a position to create and control the necessary communications traffic and facilities within all the other nations with which it expects to communicate. These international arrangements, covering many details of system finances, facilities, operations, and utilization, are likely to present legal questions which cannot be settled purely by reference to national space law, new or old. Such questions are but another illustration of the need for a second kind of space law, one that is more than national.

The second kind of space law is international—space law produced not by a single nation but by agreement among two or more nations. Such an agreement may be in written form, as in the case of treaties, conventions, or United Nations actions, or it may be just a common understanding, developed over a period of time somewhat like a custom, in which case its existence and precise terms may be somewhat harder to determine. Like national space law, international space law may be newly created for space, or old law that can be applied to space.

International space law, like most international law, has only part of the usual institutional machinery (legislature, courts, police force) familiar to every citizen under national law. These shortcomings in the way international law is created and applied have existed for many centuries, and at times have seemed to result in total breakdowns. This has led some persons to feel that international law does not really exist as true law. A similar feeling probably affected some of the early Western settlers in the rough, "lawless" frontier communities.

Actually, international law is very real, and does affect the conduct of nations to a considerable extent. Even the outbreak of a war does not cause the disappearance of international law, which has definite rules covering the treatment of prisoners and enemy aliens and the rights of neutrals and belligerents. Anyone who doubts the importance of international law in both peace and war should realize that it is considered to merit serious practical study by men other than scholars—not just diplomats but also, for instance, naval officers, businessmen engaged in foreign trade, and investors in foreign enterprises. This is true even though the machinery for creating and enforcing international law may be incomplete. In space, as in the early West, there can be law even where there is no full-time sheriff, as long as there is a definite and resolute community understanding. The presence of a possible common danger—whether from cattle rustlers or from the abuse of space power—is a great stimulus to developing such understanding, and many frontier communities and other human groups have organized themselves quickly and without outside help to provide some form of law and order.

Today, a degree of understanding about space already exists in the community of nations. There seems to be a general consensus that space should be explored and used, and that it should not become the exclusive province of any one nation. These views were reflected in a resolution passed by the General Assembly of the United Nations with the support of both the United States and the Soviet Union in 1961, which recognized the common interest of mankind in furthering the peaceful uses of outer space. While such a consensus may be far short of constituting a developed body of space law, it is a good foundation.

At the present time few questions in international space law can be answered with certainty. One cannot even be sure that we yet know which questions will prove important. Nevertheless, there are several interesting questions which can fruitfully be discussed.

Perhaps the first question to arouse public curiosity about space law was, "What is the relationship between space and national sovereignty?" The background against which this question has usually been asked is the well-known fact that national sovereignty is universally recognized to include the superadjacent airspace. In other words, under international law each nation has the legal right to exclude aircraft or other objects from entering the airspace over its territory, except as it may have relinquished some of this right. National sovereignty in airspace, like the right of a nation to exclude subsurface entry by tunneling or slant drilling, follows from the traditional concept of territorial sovereignty, which has in the past been described as giving the sovereign the same right to exclude others from his domain, down to the center of the earth and up to the sky, as he enjoys on the surface. A similar doctrine as to the rights of private landowners has long existed in Anglo-American common law.

However, somewhat as the rights of a landowner are no longer viewed as barring the ordinary movement of an airplane over his land without his consent, so the sovereignty of nations has not been seriously asserted against the orbital movement of satellites over national territory. While there is no formal treaty on this point, there seems to be a definite understanding among nations that national sovereignty, although it reaches upward at least as far as conventional aircraft can fly, does not reach to the level at which artificial satellites can maintain normal orbits, or beyond.

A little reflection upon simple astronomy makes it easy to see why the concept of territorial sovereignty has not proved well suited for the extension of national boundaries up through space. The daily rotation of the earth on its axis is a motion not maintained by objects in space, as shown by the fact that most objects in space appear from the earth to rise and set. Therefore objects in space, including distant galaxies, the sun, planets, the moon, and artificial satellites, would be crossing the boundaries of various nations several times each day if these boundaries continued upward into space. The man in the moon would need a lot of passports and visas! An exception to this would be a so-called stationary satellite in equatorial orbit, but this is a relatively minor exception measured against all the natural and artificial space objects which inevitably pass over many countries each day.

The apparent worldwide recognition that national sovereignty stops before penetrating space does not seem in itself to pose insuperable problems for the safety or dignity of any nation. If space, like the high seas, is beyond national boundaries, this does not mean that space will be beyond applicable international law. A nation may take positive action consistent with international law against activities on the high seas that violate that law, such as piracy, despite its lack of legal power generally to exclude others from the high seas. Similar national rights, short of sovereignty, may come to be recognized in space. For instance, a weapon of mass destruction, such as a nuclear bomb placed in orbit and remotely controlled for deorbit and detonation on a target in some other country—an operation which Soviet spokesmen are reported as having declared to be within their capability—might come to be regarded as an active menace in terms of international law to the nations thus threatened. A treaty would be the clearest way of covering this subject in the law of nations.

To many people, the concept that national sovereignty includes airspace but not "space" suggests the further question, "Where is the boundary?" The boundary question is complicated by the notion that there may be an intermediate zone, above the airspace but below the level at which satellites can successfully continue to orbit, a sort of buffer region where neither conventional aircraft nor satellites can operate for long. The problem is further complicated because the ultimate upper limits of practical aircraft operation are not known and agreed upon, and because of the possible development of hybrid aircraft-space vehicles.

For such reasons, it may be a long time before a precise boundary, like the three-mile limit at sea, is recognized between national sovereignty and space. But this should not be cause for pessimism or concern. After all, few landowners today know precisely how far up above the surface their rights as landowners continue, and by and large neither they nor society seem to suffer from the lack of precision here. If space law develops soundly to encourage the peaceful uses of space, if it adequately recognizes the rights of nations whose interests may be affected by the space activities of others, and if it deals effectively with miscellaneous space problems, the task of securing agreement on a precise airspace-space boundary may not be sufficiently important to warrant the effort. Once it is recognized that space is more or less free under international law for peaceful uses generally, the more important question may be, "To what extent should these uses be regulated for the safety and benefit of mankind?"

The question whether space activities should be regulated or controlled by an international authority of some kind in the interest of sound scientific progress and world public safety is basically more of a policy question than a legal one. It calls for the ideas of lawyers, scientists, and all thoughtful citizens, as well as statesmen in office. The wisdom of such regulation may change considerably in time. Today, the real question may be whether the optimum exploration and the safe utilization of space for the benefit of mankind demand actual control by a permanent international body, so as to warrant every effort to create such a body, or whether some other arrangement might be preferable, at least at present.

On this issue of control, the history of science certainly tends to demonstrate that scientific progress is greatest when scientists are free to undertake experiments without much outside control, and when they can readily exchange their theories and data. An international body with the power to approve or disapprove proposed experiments could, if that power were arbitrarily exercised, impede scientific progress, according to some scientists here and abroad. This suggests that, rather than an international control center for space exploration and experimentation, attention might first be given to better international coordination and consultation, short of attempted control.

Moreover, the United States, as a free society, might well hesitate to recognize a general international control of space traffic and activities, at least without real safeguards against abuse of power. This country has always recognized freedom of movement for its people and their vehicles. Our philosophy of terrestrial traffic control is not grounded on an absolute right to exclude citizens from public highways, waterways, or airways, but only to regulate traffic to further safe and expeditious movement and to protect the public interest. We would find it very distasteful to deal with traffic congestion simply by ordering part of the traffic off the highways, or to deal with traffic hazards simply by ruling off the roads all who statistically may be below-average risks. This balanced philosophy of traffic control, together with pertinent technical facts peculiar to space, would presumably influence our policy on traffic control for space—if, as, and when there appears to be a need for such control.

There are, of course, actual or potential space activities which are not intrinsically innocent or beneficial to mankind, and may well prove harmful. To control or terminate such activities would be in the interest of the United States and the world. Where this is the case, the United States is anxious to move forward toward specific international arrangements designed effectively to curb such activities. Our long efforts to arrive at a nuclear test ban, which would include nuclear tests in space, demonstrate our interest in achieving a rule of law in space for activities of this nature.

We do not, however, classify non-aggressive military uses of space as harmful or otherwise impermissible. We do not, in supporting the peaceful uses of space, intend to exclude such nonaggressive activities as observation. By "peaceful" we mean nonaggressive, not necessarily nonmilitary. In our view, a Navy hospital ship, or a harbor defense net, is much more peaceful than a civilian throwing a bomb.

The use of space for travel or other peaceful activities raises the question of legal redress for injuries or damage that may result. Where redress is sought for injury caused by the space activities of one's own government, national law can be invoked, or if necessary expanded, to assure prompt and fair recompense to innocent victims. Where redress is sought against a foreign government, the injured person's own government would normally present the claim through diplomatic channels, relying on precedents under international law for the satisfaction of injury claims.

It is to be hoped that all nations which are active in space will maintain the highest standards of safety in order to minimize the risk of injuries to persons and property, but that if such injuries occur, compensation will be paid where justice so requires, in time to minimize hardships. In cases where the legal right to such compensation as a matter of law may not be clear, perhaps compensation should be granted as a matter of national space policy.

As those who ponder the legal problems of space move far enough in their imagination to reach the moon or a planet, the question of national sovereignty or other national rights is likely to reappear, not as a projection upward of terrestrial boundaries but as an incident of discovery and occupation. Terrestrial law presents at least two different analogies: the older tradition of national claims, based on discovery and occupation as developed during the age of Western colonization, and the more recent pattern offered by the international treaty on Antarctica, in which the signatory nations, including both the Soviet Union and the United States, renounced claims to that continent and internationalized it. It is impossible to say at this time what the precise legal status of celestial bodies will become in the future, but under a resolution adopted by the United Nations General Assembly in 1961 no nation is entitled to recognition by other nations of any assertion of sovereign rights to the moon or other celestial body.

Perhaps a new hybrid principle will be evolved for celestial bodies, emphasizing not sovereignty within boundaries, but primary rights of a nation in a localized facility created by its own efforts. In addition, some facilities might result from joint efforts by two or more nations with partnership rights. Such principles might make possible the cooperative development of celestial bodies and at the same time might conceivably further a pattern of international competition on earth that would be basically constructive rather than destructive—the kind of competition we try to encourage within our own society.

These hopes are not based solely on the example of Antarctica. No nation owns the oceans, and no nation could successfully claim sovereignty over an ocean by anchoring a vessel in the middle of it, even for a long period of time. Nevertheless, any nation which did so, or which erected a tower or artificial island on the high seas or the continental shelf, would presumably have primary rights in its own ship, tower, or island. Whether such analogies may prove useful is largely dependent on the future course of space exploration, as well as on principles of ethics and common sense, factors which hopefully will play a significant role as international space law develops.

We also may hope that the development of international space law will be concerned not only with the rights of nations, but also with the rights of individuals. Today, United States citizens are free to sail on the high seas, engage in deep-sea fishing, and conduct other activities in areas beyond our territorial sovereignty, subject of course to United States law, international law, and the law of other countries in cases where such law applies. Should it ever become economically feasible for private citizens to undertake trips to the moon or planets and while there to engage, for instance, in mining or other productive activity, a way should be found by the United States and by the community of nations to authorize and regulate such expeditions under appropriate terms and conditions. After all, in any society, individual initiative is likely to remain an important ingredient of human progress.

Perhaps the most intriguing question posed by those who ponder the space law of the future is, "What rules should be provided for dealing with intelligent forms of life from other planets whom we may encounter?" Even on a question like this, our past experience does not leave us wholly without guidance. Assuming that these beings are not so hostile and dangerous as to force us into a defensive position, one would hope that our attitude toward them would be humane and civilized. The ancient policies of conquest, exploitation, forcible conversion, and colonization which victorious nations historically imposed on weaker peoples should certainly not be followed; instead we should try to emphasize establishing communication, mutually beneficial trade, friendly cultural understanding, and travel on an exchange basis. Think what a fascinating dinner guest or golf companion we might find in an intelligent creature with six eyes, wings, and a pouch like a kangaroo to carry his wife in! Of course, this assumes that technical problems such as language differences, varying immunities to contagious diseases, etc., will not prove insuperable.

The future growth of space law in most respects will depend on the particular uses to which space will be put. This in turn will chiefly depend on science, technology, economics, and politics. Questions which have loomed large up to now, such as the upper limit of sovereignty, may in the long run fade into lesser importance with the development of a meaningful and mature body of space law. Such a body of law will probably be oriented to particular space activities, with abstract legal concepts in a subordinate role.

Critics have said that nations are moving too slowly in establishing international rules and regulations for space. But premature or unwise laws, like the absence of needed law, could prejudice the potential benefits of the Space Age. For this reason care must be exercised to see that space law is developed at the right time and in the right direction to encourage and not impair promising space activities, while deterring harmful activities and harmful features of worthwhile ones.

The responsibility for such constructive and timely development rests primarily with the world's statesmen and legal specialists, serving not only in national governments but also in the United Nations and related international organizations. These men, however, are likely to be more effective if they receive the intelligent support of informed citizens and especially educators, both the academic and the journalistic variety. These latter groups may be able to help infuse into expert and popular opinion on space law those broader perspectives which often prove so valuable in arriving at sound legal policy.

Should some new turn in world affairs or in space activities suddenly present an important space law question for which there is no specific answer in existing codes, treaties, or treatises, there need be no panicky fear that the rule of law will on that account be abandoned in favor of sheer power. As legal history shows, many important questions have arisen in the past for which the answer had to be worked out rather than merely looked up. In such cases, the guides to solution are frequently found in the factual nature of the subject to which the question pertains, and in whatever analogies existing law may suggest.

However, the facts and available legal analogies do not always dictate the nature of final solutions. Sometimes equally important are the character and background of the persons charged with obtaining the solutions. This is not to say that the rule of law yields to the rule of men. The rule of law depends less on a supply of predetermined answers for possible questions than on a firm inclination toward fairness, orderliness, and wisdom in the adjustment of human relationships. If the persons in power are men who see their nation's self-interest narrowly, their solutions are likely to be opportunistic, unstable, and a source of more problems than they settle. It must be recognized that such attitudes have often seemed predominant in international affairs, and their presence may account for the fact that the rule of law in international relations is by no means accepted to the degree that it prevails within most countries. On the other hand, if nations are served by men who bring to the problems facing them not only technical competence but also a searching imagination, ethical integrity, a real desire to recognize the equities of others, and an ingrained conviction that the rule of law in human relationships is essential to the survival of civilization, the chances are bright that mankind will successfully surmount the inevitable crises in the future growth of space law.

ATOMIC POWER: THE KEY TO SUPREMACY IN SPACE

Glenn T. Seaborg

The two leading and competing world powers, the United States and the Soviet Union, have launched a new era—the Space Age—enabling man to escape the bonds of gravity that have, until now, confined his endless search for new knowledge and new frontiers to the planet Earth. The Russians hurdled the barriers to space exploration first because they began a large and intensive program toward that goal before we did. The result is that, as of today, the USSR has a head start in "spectacular" space firsts.

The adventure, however, has just begun, and its outcome will be determined by how far man can explore the vast reaches of the universe; and to these ends, the infinitesimal atom holds the key to the future. Indeed, success in space ultimately will be won by that nation which first harnesses the energy of the atom to carry men and machines to Venus, Mars, and beyond.

In this sphere, the United States has attained a first in space which has not thus far captured public attention. This was achieved on June 29, 1961, with the successful launching of the Navy satellite, TRANSIT IV-A, the first satellite to carry out its mission by means of nuclear power. TRANSIT IV-A carried into orbit instrumented space beacons designed to transmit navigational data to specified tracking stations on earth. An "atomic battery," weighing less than five pounds, was one of the power sources for the space beacons. This tiny battery is designed to operate for five years, although it contains enough atomic fuel for many more years. The "atomic battery" is the symbol of an inevitable and most promising marriage between the microworld of the atom and the enormous vastness of space.

The application of nuclear energy for use in space exploration is not a new concept. In 1919, Dr. Robert H. Goddard, this nation's great space pioneer, lamented that nuclear energy was not at his command for his rocket experiments. He said, ". . . we are at once limited since subatomic energy (nuclear energy) is not available . . ." It was not until 1942, twenty-three years later, that this awesome energy of the microworld for which Dr. Goddard had yearned became available. At that time, the late Enrico Fermi and a group of researchers achieved control of the fissioning nucleus of the atom. Preceding the Space Age by fifteen years, this achievement has made available a source of power which removes the limits Dr. Goddard feared would otherwise be imposed upon man's advancement into deep space.

It was 1955, two years before the launching of Soviet Sputnik I, that work began on a nuclear rocket program designated Project Rover. To date, national investment in Project Rover is well over $300 million; and many more millions of dollars will be required before an operational nuclear rocket engine is developed. But this is essential spending if we are to attain our national objectives in space.

These objectives go far beyond landing men on the moon and returning them to earth—our priority goal in space for the 1960's. Beyond the moon are other planets and the deeper reaches of space which await exploration. An examination of the energy or power requirements for these missions indicates why the investment of time, mind, and public funds to develop nuclear power systems for space is justified. To get a spaceship from Earth to Mars and back again demands reliable propulsion of tremendous power that can be generated at command to propel the ship toward Mars, away from the earth's gravitational field. Power also is needed to maintain life-support, communication and other vital systems within the vehicle. In addition, there must be power to propel the vehicle and its crew safely back to earth. Weight and space limitations make it essential that the power be available in the form of high-energy output at low weight with guaranteed long-term reliable performance. Only nuclear-energy systems can provide such power reasonably and economically.

While, at the present time, no one would be tempted to abandon chemical propulsion systems for launching vehicles into space, they cannot fulfill these requirements as well as nuclear energy. Present chemical rocket systems are limited by the energies that are released in the combustion process. The combustion energies provide the "specific impulse" or thrust power —the pounds of thrust per second for every pound of fuel used. Specific impulse is a unit of measure of energy output for rocket systems roughly comparable to miles per gallon of gasoline for an automobile. The higher the specific impulse, the more efficient the fuel and the rocket engine, just as the more mileage obtained from a gallon of gas, the more efficient the gasoline and the engine system of an automobile. The present kerosene-oxygen-fueled rockets such as the Atlas and Titan have specific impulses of 300 per second. (For every pound of propellant, 300 pounds of thrust are produced per second.) Hydrogen burned with oxygen will provide the combustion energy in the upper stages of the advanced Saturn rockets designed for our manned lunar launchings. The specific impulse of this hydrogen-oxygen combustion is about 400 per second. A Saturn I booster with a hydrogen-oxygen-fueled upper stage will have the thrust power to put a payload of approximately 11 tons in a low earth orbit. This is about 50 per cent more than the weight-lifting capability of a Saturn I with a kerosene-oxygen second stage. As you can see, an increase in specific impulse results in a proportionately much greater increase in weight-lifting capability.

Present research indicates that a hydrogen-fluorine-fueled combustion system is potentially the most efficient chemical system possible. The specific impulse of such a system is 450 per second, and this appears to be the attainable upper limit for chemical fuels. To perform long-range manned missions beyond the moon economically, the specific impulse must be increased well beyond this level.

Specific impulse depends, among other things, on the propellant temperature and the molecular weight of the propellant. The higher the temperature and the lower the molecular weight of the fuel used, the greater the specific impulse. Because it has the lowest molecular weight, hydrogen is the most promising rocket propellant known. However, in a chemical combustion system, oxygen or some oxidizer must be used in a combination with hydrogen to permit the necessary chemical reaction to occur and release the energy to heat the combustion products. In a nuclear reactor, any fluid can be heated without an oxidizer to extremely high temperatures—above 3,500°F. This has been a major factor in stimulating interest in developing a nuclear rocket. Research has demonstrated that by using hydrogen in a nuclear reactor, a specific impulse of about 800 and more per second can be obtained—an energy output roughly double the maximum obtainable from the most efficient chemical system. Thus, in a nuclear rocket a pound of propellant will yield a thrust per second about twice that from a pound of the same fuel in a chemical rocket without the weight burden of an oxidizing agent.

The elimination of an oxidizing agent also eliminates some of the complex systems of valves, pumps, and tanks associated with the oxidizer part of a chemical system. However, a nuclear reactor does bring along with it additional control systems to insure its safe, reliable operation.

The substantial reduction in vehicle weight by increasing the efficiency of the rocket system is the major advantage of a nuclear rocket. How substantial is the reduction? It has been estimated that a chemical rocket system for a manned round-trip to Mars might weight about 10,000,000 pounds—the equivalent of 625 San Francisco cable cars—in an earth orbit. It would, for example, take about 100 separate launchings with the advanced Saturn booster* to put such a heavy payload into an earth orbit. The problem of assembling the parts and transferring the necessary quantity of fuel and oxidizer into the vehicle for the Mars trip would be fantastically complex. A nuclear system for the same trip would weigh only one-fifth or one-tenth as much, reducing the launching problems substantially and simplifying somewhat the assembly process. While the cost of such a system is considerable, the efficiency to be gained and the wide expansion of scientific exploration certainly justify the investment.

As I mentioned earlier, Project Rover is this nation's nuclear rocket program. The foundation of Project Rover is the experimental reactor program, appropriately named KIWI for the nonflying bird of the same name; for the KIWI reactor will never fly. These reactors are being ground-tested in order to determine the modifications and design configuration necessary to develop a reactor of the required size and shape to fit into an engine system that will fly. NERVA (Nuclear Engine for Rocket Vehicle Application), the next step, is already under way. This stage of development is equivalent to the static engine testing of a chemical rocket system and will be followed by a flight test phase known as RIFT (Reactor in Flight Test).

Project Rover is managed jointly by AEC and NASA through the operation of the Space Nuclear Propulsion Office (SNPO). AEC is responsible for reactor development; NASA directs vehicle design. Incidentally, the administration of all the facets and stages of Project Rover by SNPO is an example of a coordination of effort and diverse function by industry, government, universities, and research centers essential today for the advancement of science and technology for public benefit. The Los

* The advanced Saturn will be able to put about 90,000 pounds in orbit.

Alamos Scientific Laboratory, which is operated for the AEC by the University of California, does the work on KIWI. Aerojet General Corporation and the Westinghouse Electric Corporation are the main contractors for NERVA. The Lockheed Missile and Space Company is under contract to the NASA Marshall Space Flight Center for RIFT, while technical support for engine design and development is provided by the NASA Lewis Research Center.

The objective of Project Rover—to develop reliable nuclear rockets for propulsion for long voyages—is only one aspect of the application of nuclear energy for space. Power is needed for electric propulsion systems, to maintain and operate manned space stations, to establish bases on planets, to transmit data for meteorological, navigational, and communication satellite systems, to assure functioning of life-support systems and a variety of other operations. Compact electric generators using radioisotopes or nuclear reactors appear promising as a source of power for these tasks. Each generator can be built for a specific power requirement ranging from a few watts to megawatts of electricity. To produce these nuclear-fueled generators, the AEC has undertaken Project SNAP (Systems for Nuclear Auxiliary Power).

Our first nuclear space success, the still functioning "atomic battery" in TRANSIT IV-A, was a SNAP device using radioisotope plutonium-238 as a fuel. The heat energy from the alpha* decay of the radioactive plutonium was converted into 2.7 watts of steady electrical power, the amount needed for the TRANSIT space beacons to transmit data enabling mariners to get a precise "fix" on their positions at sea at any time of day or night. Other SNAP isotope power units or "atomic batteries" only slightly larger and heavier and capable of delivering ten or more times as much power now are under construction for a variety of satellites used for space probes, lunar examinations,

* There are three types of radiation: alpha, beta, and gamma. Alpha is the least penetrating and, therefore, the easiest to handle for power production.

and other applications.

The principles underlying isotope power units are simple. Heat is generated by the decay of the radioisotope and absorbed in the surrounding material; and by means of a suitable energy-conversion device, either thermoelectric or thermionic, the heat is partially converted into electricity with the remaining heat dissipated into the external environment. Isotope units can generate up to several hundred watts of electric power; and since the heat produced by the radioactive decay can be directly converted into useful electricity, there are no moving parts to wear out and no maintenance to worry about. Long-lived and reliable, they are ideal for use in satellites since they are impervious to the expected conditions and hazards in space. They also are suited for installation in remote, unattended land- and sea-based installations. In fact, radioisotope units installed in Navy and Weather Bureau stations in the Arctic and the Antarctic have been operating for nearly two years without maintenance.

Radioisotope-fueled generators can deliver power levels of the order of tens and hundreds of watts; but nuclear reactors are required for the kilowatts and megawatts of power needed for manned space stations, bases on planets, and communication satellites for direct space to home transmission of television broadcasts. Several SNAP reactor systems are being developed, the smallest designed to provide 500 watts of electricity in continuous operation for at least one year and the largest to provide from 300 kilowatts to 1 megawatt of electrical output for more than a year. The 500-watt SNAP reactor proper, without the auxiliary gear, which can be bulky, is about the size of a five-gallon container; and the megawatt reactor proper, which will produce far more power, will be only very slightly larger. Considering that the average house requires about 1 kilowatt of electricity, you can see that these reactors will be capable of providing power on a continuous basis ranging in amounts sufficient for half a home to that required for a thousand homes.

The smallest SNAP reactor, the 500-watt system, is designed for use in Air Force satellites for orbits up to a year. The megawatt reactor, now in the preliminary phase for system design and fuel and component development, shows great promise as a power source adequate for communications satellites capable of transmitting television programs directly to our homes rather than requiring receivers to amplify and relay messages as in the Telstar and Relay satellites. Such a direct transmission system could be placed in an orbit with an altitude of approximately 22,000 miles. At this altitude the period of revolution for the satellite would coincide with the 24-hour period of the earth's turning. The satellite thus would appear stationary over one spot on earth. Three such satellites properly distributed could, according to present estimates, cover transmission of television and radio for the entire earth. Only a large SNAP reactor could provide the power demands of such a world-wide communications system.

With this exciting potential—present and foreseeable—of the marriage of atom and space, not only for removing the barriers to man's investigation of the universe but for bettering life on earth, it is startling to realize that mankind is still only on the verge of the Space Age. In a short time much has been achieved and the promise for the future is great, but much more remains to be done and great problems lie in the way.

A major problem is the development of materials not now available which are essential for nuclear power in space. Our nation must undertake a much larger program of research and development in materials. Only such expanded research can provide boiling liquid metals for efficient transfer of heat, lightweight shielding to protect against radiation exposure, more high-temperature nuclear-fuel materials, and a host of new structural materials for future high-power lightweight systems. Other unknowns exist. However, the promising performance of our early research and development in SNAP and Project Rover, even in the face of these problems, justifies—indeed, dictates—our most vigorous pursuit of the goal of nuclear power in space.

In any discussion of the promise of nuclear power, the question of peril inevitably arises. What about the hazards of radiation? Can Rover be flown safely? What happens to the radiation in SNAP isotope units not converted into electricity? Will it contaminate the environment into which it escapes or can it be safely contained? A very important part, and properly so, of the Atomic Energy Commission's statutory responsibility in the space program is safety—safety in flight, in maintenance, in ground operations, in re-entry mechanisms. Some of the potential hazards in the nuclear rocket program and in the use of nuclear power packages are the same ones we have met and have learned to manage in other applications of nuclear energy. For example, there are over one hundred nuclear reactors of all types in operation in this country today. There are also some new and unique situations to be faced in space. These include consideration of the maximum conceivable accidents that might arise in the launching of a rocket with a nuclear upper stage or a vehicle containing a large SNAP system. What are the hazards of the malfunction or abort of a nuclear rocket and its subsequent re-entry into the earth's atmosphere? And, finally, the ultimate disposition of the nuclear engines and power units in space must be considered. These are tough questions. Not all the answers are readily available; but enough are to justify our going ahead with nuclear power for space in a spirit of cautious and watchful optimism.

First of all, a nuclear rocket *will not* be launched from the ground. It will be boosted by a chemical rocket out of the earth's atmosphere and will become operative at command, either built-in or given from earth, only when it is at a level in space where the radioactive debris or fallout discharged in its operation could pose no hazard to earth and its environment. A nuclear reactor is not an atomic bomb and will not explode like one if an abort should occur and it should crash to the ground. However, prior to any launch, extensive tests will be made to obtain safety information. For in applications for space, as in other nuclear programs, a conservative approach has been taken toward safety and will be continued until sufficient experience is available to provide a more accurate basis for judgment. This approach involves extensive safeguards to prevent accidents and also substantial efforts to minimize the consequences of accidents in case the precautions taken to prevent them should somehow fail.

Much of what is within our grasp depends on the further development of fundamental knowledge and upon the training of larger numbers of scientists and engineers. Past experience shows that success in this endeavor can largely be measured by the involvement of universities and colleges with government and industry. The Atomic Energy Commission early set an example in this direction by bringing the universities into atomic research thereby giving the nation strong programs and basic research and a large pool of trained personnel in the nuclear field. The need is just as great for similar bonds between the universities and space technology. NASA is meeting this need with its education support program, which has broadened university participation while, at the same time, making it possible for the schools to maintain independence of thought and research. Success in space has thus far largely been measured by the feat of our astronauts. But while we continue to thrill to their heroic efforts and pioneer achievements, we should also be aware of exciting developments in education, generated by the scientific and technological advances in our space program.

There has been a lot of discussion about the accent on space. Questions have been raised as to whether the accent is not too heavy. My answer to this is that we have a pioneering heritage. Our forebears were adventurous people who broke with old ties to come to the New World, who conquered a new continent, and who have continued these traditions in science, industry, and technology. When it was possible to explore the atom, we did not hesitate. It now has become possible to explore space. We dare not shirk the adventure. We cannot draw a curtain over a New World that is within our grasp. We cannot sit at home, so to speak, and hear second-hand of new wonders that men have pondered through the ages. Our enthusiastic participation on the frontier, wherever the frontier exists, is necessary for our continuation as a dynamic and creative people. If there were no other reasons for space exploration—and there are a great many more—this one would be good enough for me.

CAREER OPPORTUNITIES IN THE SPACE AGE

R. W. Retterer

The word *career* implies ever-increasing opportunities and the challenge of ever-greater responsibilities for the individual; and nowhere is there more opportunity and challenge for the career-minded person than on a new frontier. Most rewarding careers of the past have been related, in some way, to the frontiers faced by the human race. This is no less true today, at the dawn of the Space Age, than it was two hundred, five hundred, or a thousand years ago.

The nature of the frontier changes with time and human effort. Each new frontier sets and defines its own talent requirements. Young people today, as well as their parents and educators, would do well to assess the talent requirements of our present frontiers. These requirements differ markedly from those of only a generation ago; and unless the differences are understood, we run the serious risk in decades to come of career positions looking in vain for men while men go looking in vain for careers.

Yesterday's frontiers, for the most part, were physical; and yesterday's careers required physical labor. It took men of strong back, strong will, and immense courage to push back the wilderness and forge great civilizations. But today's frontier does not start at the edge of a forest. Rather, it begins just a few miles above our rooftops in outer space. The conquest of space cannot be achieved by physical strength, determination, and courage alone. Its conquest will depend on the intelligence and the creative power of the human mind, the advanced tools conceived by that mind, and the skills which enable him to use those tools.

One major fact should be crystal clear to anyone assessing career opportunities: the Space Age is a product of rapidly advancing science and technology. Science, technology and space are almost synonymous. They have entered just about every aspect of our lives. It is quite reasonable to assume that the careers of tomorrow, in virtually all fields of endeavor, will be linked intimately with science and the changes which accompany its advance. Some of those changes are immediately evident in the host of exciting career opportunities that present and future generations may pursue. Less than a decade ago there was no such person as an astronautics engineer. There was no such person as a biophysicist. There was no field called operations research in which people employ the most advanced analytical tools in order to find the best way of performing highly complex tasks. There was no branch of medicine called aerospace medicine; no such field of endeavor as microelectronics. The army of men who now make, sell and service electronic computers did not exist. When this year's high-school sophomore was born, there was not more than a handful of computer programmers. The electronic computer itself was a laboratory curiosity. Today there are more than 40,000 computer programmers in the United States alone, and by 1970 America will need close to a quarter-million!

While not all the careers of the Space Age will require college training, it must be made clear that almost all will require some level of specialized skill. In any attempt to rank careers, the Ph.D. scientist and engineer probably will continue to enjoy the top positions. They are the professionals who will be in greatest demand, for they have been trained to force nature to reveal her many guarded secrets. Once these secrets are available they offer many opportunities for practical application. Thus, for each Ph.D., five to ten engineers at the B.S. level can be employed, and ten to fifteen skilled workers below that. Without the top level of skill, men of lesser skills will lose their opportunity.

Science in general is a good bet for any career-minded person who exhibits the required talent and interest. More than 90 percent of all scientists who ever lived are alive and working today. That percentage will grow during the next few years. Today there are approximately 875,000 scientists and engineers working in industry; 125,000 in government; another 110,000 in colleges and universities, and 60,000 at independent research laboratories. Conservative estimates indicate that by 1980 we will require at least twice as many as will be available if the present rate of increase is not substantially enlarged.

There is an annual demand now for 72,000 new engineers and scientists. Our schools are producing only about 35,000 technical graduates a year. The Soviets, by contrast, are graduating over 100,000 technical people every year, and there is reason to believe that there is considerable quality as well as quantity.

Naturally, the minimum standard for most creative careers in Space Age science and technology is a good education. The desired level is the doctor's or master's degree, and there is good reason for such high standards. While the fundamentals of science are taught and secured at the lower levels of education, it is at the graduate level that the greatest productivity of the mind is achieved. In graduate work, a certain amount of "channeled interest"—call it specialization, if you will—is imparted to the would-be creative scientist. This channeled interest is very necessary in the Space Age, for today no man can say, as did Sir Francis Bacon 350 years ago: "I take all knowledge for my province." The complex nature of scientific knowledge since mid-century precludes such lofty ambitions.

The need for new Ph.D.'s receiving degrees in science and technology each year is estimated at about 100 per million population. This means our educational system should produce roughly 19,000 Ph.D.'s a year if we are to take full advantage of our science-oriented economy. The sad fact is that we are graduating only 12,000 doctorates a year in *all* fields, and only 30 percent of them are in the physical sciences and mathematics. This production must be multiplied by five, without sacrificing standards, if we are to meet vital national commitments in the Space Age.

The Ph.D. situation in mathematics is particularly grim. There is a critical need in our society for advanced mathematicians. These are the men and women who are perhaps best equipped to analyze abstract relationships, ferret out the necessary from the unnecessary, and define broad and meaningful problems. Yet our educational system provides only a handful of Ph.D. mathematicians each year.

One might think that there is a shortage of people capable of proceeding to the Ph.D. level, but this is not so. The fact is that more than 75,000 high school graduates annually exceed the median level of IQ and creativity indices for individuals now receiving the Ph.D. Somewhere along the line we lose them to other pursuits. As Dr. Lloyd V. Berkner so aptly put it recently: "To talk of exhausting the supply of qualified individuals is nonsense. The job is to enlarge opportunity for their training and to identify these students, to motivate and to support them."

Talent in creative science and technology, of course, is far from being the only talent the Space Age demands. There will continue to be a great need throughout government and industry for trained managers. Management today requires the ability to lead people into productive channels. A few years back it might have been possible for the boss' son, having toured the tennis courts of Europe, to come home and step into his father's shoes. Today even the boss' son requires rigorous training. Managers in the Space Age must understand the new tools of production, computers, and economics. Without sufficiently skilled people capable of closing the gap between human goals and abstract science, those goals may never be realized.

The social scientist in the Space Age has a special responsibility to see that material, intellectual, and cultural values accrue to society as a whole. Problems and dislocations are bound to occur, just as they always have in the past when labor-saving and more productive tools have come upon the scene. The psychologists, sociologists, social workers, and politicians must find ways to ease the transition from a society largely dependent upon physical labor to one in need of mental labor and skills. They must find ways to impart a sense of usefulness to workers on every level. They must find ways to match men with available jobs through training, retraining, and relocation, if necessary. Another key job for the careerists in the social sciences will be to set forth guidelines that will permit us to make the most productive and rewarding use of increased leisure.

Indeed, just about every traditional career involving the human intellect rather than human muscle is taking on new responsibilities peculiar to the Space Age. Sociologists and psychologists must find new ways to motivate individuals toward productive careers, and new ways to train people. A branch of psychology that has grown up only since the start of the Space Age offers a unique and exciting career blend of psychology and physical science. It is called human factors engineering. Those thus engaged seek to find the best ways to permit man and machine to get along with each other. It is they who must, for example, determine how many switches an astronaut can safely handle in a space vehicle, where they should be placed for maximum effectiveness, what color they should be, how they should be operated, in what sequence—and then help the astronaut learn how to use them properly.

Lawyers and political scientists must produce the appropriate governmental and legislative foundations that will permit our goals to be reached in the shortest possible time. Lawyers have the added responsibility of finding answers to international and interspatial jurisdictional problems which already have arisen as man penetrates space with increasing frequency.

Those who have a strong affinity for language will be needed to communicate the meaning of space and the Space Age to the public at large. Reporters must be around to report. Technical writers will have to prepare the documents and manuals that invariably accompany technical "hardware." Language specialists will be called upon to perfect the means for machine translation of ideas from one language to another, and someday will be asked to develop the means for communicating with extraterrestrial life that most certainly exists on other planets in other corners of the universe.

In the Space Age, the humanities and arts also take on new meaning and new responsibilities. Not even the scientist in the loftiest laboratory can shut the door on the past, or on the rest of society. He must acquire the polish of an educated man above and beyond professional competence in his chosen career. To help him achieve this there must be many educators, artists, entertainers, writers, poets—careers which can be as rewarding as any. Moreover, the scientist as a citizen must be able to live with his neighbors as well as with himself. He must share those responsibilities that fall upon all who are informed. He must bring his orderly scientific mind to bear on culture, art, politics, and religion, and become a part of all those areas of human enrichment which constitute our national goals and national pride.

The Space Age clearly is the age of the intellect. While we will always need some type of service workers, such as firemen, waiters, policemen, and household workers, the outlook for other types such as farm workers, unskilled production workers, and day laborers is not particularly good. Herein lies a real challenge for our democratic society. The percentage of young people who go to college must be increased from its present 20 percent level, while the large percentage of youths who do not go to college must be prepared so that they will have a marketable skill. Just as those eligible for college must be identified early in the educational process, so must the students not suited for college be identified. The latter must be appropriately educated so that they will have special skills to contribute to society, for there is little demand for the uneducated and unskilled in today's world, and there will be less demand for them tomorrow. It does not take a college degree to become an electrician, an electronics technician, a salesman, a draftsman, or even a computer programmer in some cases. It does not take a college degree to develop from a stenographer to an administrative or executive secretary. It does not take a college degree to become a top-notch machinist, an aircraft mechanic or an auto mechanic. In all those classifications—and many more—there is a present shortage of workers and that shortage will probably continue for many years. In all of them, however, a certain degree of special skill is required, and skill means training. Instead of forcing a youngster to stay in school until he is sixteen years old, or thereabouts, and then tossing him out among the unemployed, we should make him stay in school until he learns a useful skill which is in demand.

Often we are asked what the impact of automation is or will be on the prospect for a career. Automation is the broad substitution of machines and electrical power for human muscle power and human sensory organs. In one form or another, it has been going on for centuries. Whenever American industry can find a better, more economical, more productive, and safer way of performing a task through automation, that task is likely to be automated. If automation has done anything to the career structure, it has shifted the labor demand from unskilled production workers to people who can work with their brains, and to those who have specific training, technical and clerical.

The electronic digital computer, which, rightly or not, has become the symbol of automation, has had a dramatic impact upon careers. For one thing, it has become the common denominator of all disciplines and endeavors. It is indispensable to science, technology, industry, and education. No technological advance—not even the airplane—has caught on so rapidly, or has had as great an impact upon society. The computer not only has produced careers of its own, but careers *within* careers. The scientist, banker, industrialist, law officer, civil servant, military man and small businessman—all realize some direct benefit from the computer.

The computer and the computer programmer are fundamental to our national space effort, and in advancing space exploration, they have advanced their own use and application. However, programming positions are not the only new career opportunities that have been created by the computer. The demand for programmers is matched by the computer industry's need for mathematicians to make the machines increasingly efficient; for electronics engineers to devise faster and smaller circuits and electronic memories which approach the component density of the human brain; for production experts to convert engineering drawings into realities; and for service technicians to keep the computers functioning as they are intended to function. These, and many more career opportunities, have grown up over the past fifteen years as the direct result of a single new concept—the electronic digital computer.

Recognizing the universality of application of computers in the Space Age, many universities and colleges now require both engineering and business students to complete at least one course in computer operation and programming. At Pennsylvania State University, one of many schools offering computer instruction, all freshman engineering students must take a computer course unless they are already proficient in the use of a computer. If college freshmen can learn the principles of computer operation and programming, why not high-school students? If the computer is becoming fundamental to most careers in our science-oriented society, should not its operation become part of fundamental education? This trend will continue to grow and spread to students in the social science programs, education, humanities, language, library science, political science, and economics. Computers are being used outside the academic world to make concordances of ancient documents, translate from language to language, store and retrieve complex information, and correlate vast amounts of statistical data. If one wishes to pursue a career in these areas, he would do well to master the basics of computer technology, because the computer removes much of the drudgery from man's intellectual labors. Once the drudgery is gone, creativity can assert itself in full measure.

There is no short or easy way to a rewarding career, nor is there any guarantee that a particular career field will remain stable throughout the life span of an individual. Because of this built-in uncertainty of human affairs, fundamental education cannot be over-stressed. Knowledge of fundamentals is the foundation upon which the individual can build and rebuild, train and retrain, adapt and modernize his own skills to the changing needs and challenges of the future. Mastery of fundamentals permits a person to jump from one career to another, or from an old, obsolete pursuit to a new, dynamic activity. Indeed, it might help matters in the long run if we all were conditioned to accept the idea of having more than one career in a lifetime. For example, it is not uncommon for a person who gains proficiency in a field to switch from active participation in that field to teaching, and vice versa. The men now working on lasers may have started out to be physics teachers, or rare-earth chemists, and are now actively classified among those engaged in electronics.

New and better methods of teaching fundamentals must therefore be found and applied, and these must be applied well before students reach high school. Here still another Space Age career opportunity presents itself. There will continue to be a vast need for innovators in education who can reshape curricula and reduce the twenty-two-year interval presently required for an individual to attain minimum professional competence. Other disciplines must be brought into education to improve the educational process itself. Human factors, engineering, communication techniques and management methods developed for our space program might very well be applied to better the education of our citizens. For example, television, which has been with us a long time, still has not exerted a strong, positive educational force. We are not moving fast enough in evaluating programmed learning and teaching machines. Computerized teaching machines, geared directly to the responses of the individual student, are only now being given serious trials. Advances are being made, but slowly.

We now are beginning to see in our colleges and universities new career curricula programs which cut sharply across classical disciplines. If the wonders of electronics, for example, are to be applied in medicine to benefit all mankind, then the electronics engineer must also know the human body, and the physician must know electronics. At the very least, the physician and the electronics scientist must understand each other's language and basic limitations. While we now have courses in several universities which lead to degrees in bioelectronics, biophysics, bionics and the like, there is increasing need for cross-disciplinary communication in scores of fields. Such cross-disciplinary careers truly reflect the interdependent character of the Space Age.

In order to achieve our full potential in the Space Age, we must more fully utilize the potential and skills of women. Women represent half our intellectual resources, and they must be encouraged to seek rewarding careers. One of the challenges facing parents, educators, and the social scientist today is the reorientation of public thinking with regard to women in creative careers, especially the myth that it is somehow unfeminine for a woman to enter the fields of science and technology. There are indications that we are slipping backward. A recent survey by the National Academy of Sciences National Research Council shows that women receiving doctorates in the natural sciences fell from 11 percent in 1920 to only 5 percent in 1962.

Women in science, despite their small numbers, have made very substantial contributions. Women chemists, engineers, astronomers, and mathematicians have proven every bit as capable as their male colleagues. At UNIVAC, for example, a woman holds the key post of manager of programming research and has been the guiding light in the development of several advanced computer languages. Among these is COBOL, the Common Business Oriented Language, that employs English-like words rather than complex mathematics to instruct computers.

Highly trained women are making major contributions on all

levels of industry. It has been found that the patience and manual dexterity of women technicians is extremely valuable in the assembly of electronic and mechanical equipment where tolerances must be held to small fractions of an inch. While most science-related industries, educational institutions, and the government are perfectly willing to employ women in key creative positions, not many highly qualified women are available. We must find out why and take every step necessary to change the pattern. We should not continue to limit our valuable female intellectual resources to the home, although that too is important.

Career-minded men and women of today have more opportunity than ever before in history to pursue a life's work that is exciting and rewarding, intellectually and materially. Which of the many newly opened Space Age careers the individual will choose (not forgetting older careers that have taken on new responsibilities), will depend on his native ability, how early and how well he prepares himself, the professional and parental guidance he receives, and his own personal objectives. In today's era of the intellectual frontier, the most exciting and rewarding careers will be those that bridge the realm between the known and the unknown, and those that apply newly discovered knowledge for the benefit of all mankind.

CHOOSING CAREERS FOR THE SPACE AGE

John H. Glenn, Jr.
(*as told to* Lillian Levy)

Young people who ask my advice on choosing careers often are surprised when I do not give top priority to a career as an astronaut. Certainly, by being an astronaut, you can rise high—in fact, in a literal sense, about as high as man can get; but this assumes importance only as a measure of human achievement. By this standard, the job of an astronaut—his career—as the pathfinder on the vertical frontier that is outer space does measure up as both important and privileged. Yet I would not give it top priority. Although space may be limitless, the actual number of persons who will be needed to pilot spacecraft is extremely limited, at least in the foreseeable future, largely because of economic reasons. Each manned space flight is so expensive that we cannot possibly have astronauts in the number that we have aviation pilots.

There are, however, other careers in space research and exploration that, as a measure of human achievement, are as outstanding—if less spectacular—than that of the astronaut. Before, during and after my ride in space, I was keenly aware of the fact that there was a host of Very Important Persons supporting me and upon whom the success of my career and those of the other astronauts largely depended. In the excitement of a manned orbital flight, their roles tend to be obscured since the focus is on the astronaut. His performance is under public scrutiny; and yet the excellence of his accomplishment, while a measure of his own abilities, is also a measure of the excellence of the work done by these often unsung VIPs.

Among these VIPs are, first, the scientists and engineers who devised and developed the vehicle that has enabled some of us to reach the heights of outer space, who have provided the environment within the space vehicle for human survival, and who have designed the controls that assure safe return to earth. Behind the scientists and engineers are the political leaders and officers of our government who took the required legislative and financial action in support of a national space program, the news reporters and others in communications who keep us informed and thus help promote public understanding of our national space objectives and, last but not least, all of the productive members of a free society whose taxes support our national goals in space and on earth.

This sum total and the variety of careers it represents are fundamental to our national progress; for we are not going to move forward in this country just by having a great space program—important though that is. Our advances in space must be matched on earth by great thought and action in such pursuits as the law, political science, sociology, public health, economics, religion, teaching, literature, and all the other human endeavors that contribute to the strength and richness of our modern society. If we are to progress as a nation—as the leader of the free world—we need dedicated service in a variety of careers—the best effort of all our peoples in all fields.

In choosing a career, the talent as well as the interest of the individual should be a paramount consideration. How does one discover his particular talent? In my opinion, the best way is through education; and I would, therefore, advise the career-minded individual to get the very best possible education. It is in learning that each person can best discover his talent potential.

This raises the question of whether education today is meeting the career needs of the Space Age. We undoubtedly need to increase our emphasis on mathematics and science at all levels so that the coming generation will have a better understanding of our developing technology and its effect on human affairs. However, I think it would be a serious mistake to increase emphasis on math and science at the expense of history, literature, and all the other subjects that make up our classical education. These must not be downgraded. The life of the scientist, engineer, astronaut and, indeed, of everyone is enriched by the humanities and the arts. Education which accents science and science alone will fail to meet the needs of the Space Age and could prove inadequate even for those who seek to become astronauts.

The average life-span of an American today is seventy years, of which at least fifty are potentially productive. The average career-life of an astronaut, according to present and foreseeable requirements, is from fifteen to twenty years. This leaves thirty productive years. How are these to be used? Even if his education has been limited largely to science, math, and engineering, the career opportunities of the astronaut would be secure, albeit limited to these or related fields. Certainly, he could qualify for the management of future space programs. He could teach in space-related fields. But if he has broadened his education by expanding his knowledge in other than the fields of science and engineering, his career opportunities are similarly broadened and so may be the scope of his service.

The rapid advance of knowledge today and the almost overwhelming accumulation of data require continuing vigilance that the quality of education be maintained at the highest level of excellence. But while quality is important, so is quantity.

Top priority should be given to a national effort to raise the educational opportunities for larger segments of our population. Their present lack is fundamental to the problems of delinquency and unemployment and, in large measure, is an underlying cause of the current racial unrest.

In any consideration of careers or jobs for the Space Age, it is important to keep in mind that even in space, career opportunities are virtually all "ground-based." Once we eliminate the factor of getting to a destination and back again, there are no areas of activity in space that do not or will not have an earthly counterpart. Future lunar pioneers, for example, will undoubtedly include geologists and biologists who will apply the same skills acquired for the study of the structure of the earth and its life forms to the study of the lunar surface and the life that may be present there. If we colonize the moon, we will need soil-erosion experts, behavior specialists, maintenance and repair crews, sanitation engineers, clerks, and a variety of other persons with a wide diversity of training and skills. For when man leaves earth to travel in space, he does not shed his earthly problems. They go with him wherever he goes—as do his opportunities.

To solve the problems and exploit the opportunities both on earth and in space will be a continuing challenge. We will need excellence and diversity in all our efforts, with the accent on dedicated service rather than dedication to science.

THE DOMESTIC SIDE OF SPACE: A CHALLENGE TO WOMEN

Lillian Levy

Woman's role in the United States, by custom and tradition, has been that of the homemaker responsible for the maintenance, well-being, and harmony of the household. In this domestic supporting capacity, her influence undoubtedly is substantial. As wife and mother, she is the hub around which family unity is established. As mother, hers is the dominant influence—for good or evil—upon the next generation in the molding of Presidents and playboys, social workers and social parasites, decent citizens and delinquents.

In the not too distant past, the everyday physical demands of housekeeping left a woman little time or energy for pursuits other than her domestic duties. Someone had to do the dishes, make the beds, clean, sweep, shop, cook, feed and clothe the family. All of this has been the necessary burden of sweet domesticity; and, by social convention, it was decreed and accepted as woman's work. Woman's place was assured: it was in the home. If it lacked glamour and intellectual stimulus, it had the virtue of being necessary and useful.

In the last twenty-five years, as a result of technological advances, the physical demands of housekeeping have become less onerous. Automation, in the form of such time-and labor-saving devices as dryers, dishwashers, vacuum cleaners, electric waxers, food mixers and blenders, refrigerators and freezers, has eliminated much of the drudgery from routine but necessary household chores. Another aid has been the development of new synthetic fabrics, such as orlon, nylon, dacron and a variety of plastics, which require little care or maintenance. Canned foods and processed frozen foods save hours in the kitchen, long the symbol of the housewife's daily drudgery.

A study undertaken in 1963 showed that even with all of today's labor-saving devices, the modern housewife spends a quarter of each day in the kitchen. This bondage promises to be substantially reduced by a wide variety of by-products now available from space research. These Space Age kitchen aids include pots and pans which cut down time spent in preparation and cooking of foods, new ceramic dishware that can move from freezer to oven to table, new developments in processed foods and kitchen sinks.

The time-saving pots and pans are coated with Teflon, the trade name given to a plastic material developed for rocket nose cones to protect them from exposures to extreme heat during launch and re-entry. Used on the outside of the rocket nose cones, Teflon transfers the heat away from the rocket into the air. On the inside of cooking utensils, Teflon transfers the heat more rapidly to the food being cooked. Because of Teflon's properties, the heat transfer is so efficient that a cut-up frozen chicken may be cooked thoroughly in half an hour. Thawing is unnecessary. Teflon also eliminates the need for cooking fats or oils in frying foods.

The freezer-to-oven-to-table serving dishes are made of pyroceram, a ceramic highly resistant to extremes of temperature—both hot and cold—used in the fabrication of rockets. Food can be prepared in a pyroceram dish, stored in the freezer in the same container, placed in the oven for baking and served at the table with no time wasted either in transferring foods or waiting for them to thaw.

Studies on food for space travel have contributed a wide variety of pre-cooked freeze-dehydrated foods—from shrimp cocktail to filet mignon—that require no refrigeration for storage. They can be kept safely on pantry shelves. To reconstitute them into appetizing meals, the housewife need only add the required amount of liquid and heat. Incidentally, the dehydrated foods not only are time-savers, they are space-savers as well. They are less bulky than conventional frozen-food packages.

The kitchen sink has also been improved by advances in today's technology. Ultrasonics (extremely high sound waves), which are used to clean components for the Saturn rocket for lunar exploration, are being used in experimental kitchen sinks to remove dirt and grease from cooking utensils and dishes. When such sinks are available for the home, there will be no need to scrape or scour any kitchenware or tableware; and the question of who will do the dishes will cease to be an issue disrupting family harmony.

These are only a beginning—a small part of the domestic side of space which is easing woman's work. Greater relief is promised for the woman at home by the synthetic diets now under development for interplanetary manned voyages that will take a year or more. Conventional diets are not suitable for such long-term space flights—not even freeze-dehydrated foods—because of the severe limitations on bulk and weight imposed by the propellant requirements for space travel. It takes ten pounds of a propellant to put one pound of material into space. If conventional foods were used, it would require 1,300 pounds of food per man per year—about 110 pounds per month.

NASA-supported research has already developed a synthetic diet of which one-half cubic foot in a soluble solution can satisfy the daily caloric and nutrient requirements of a man for a month; and equally important, it is also quite palatable. Adapting the synthetic diets for home use is a real possibility. Estimates are that one quart in a soluble solution can adequately feed a family of four for a week. If bulk is craved, it could be supplied by any crunchy edible foods such as cold cereals, crackers and similar foodstuffs. However, according to tests with human volunteers who lived on the synthetic diet exclusively for six months, bulk is not essential.

With synthetic diets available for home use, the family could line up for its meals at an automatic home dispenser that would pour out just the required amount in a disposable cup. In fact the cup could be a cookie confection with an edible coating to make it waterproof and thus be multi-purpose in function as container, dessert and a source of bulk. The synthetic diet could be delivered to the home as milk is today. The meal could be prepared and ready to serve in a matter of seconds. There would be no need for tablecloths, dishes, cutlery, pots, or pans —or even instruction in table manners other than a reminder not to push or fidget while waiting a turn at the home dispenser.

Another home product which may develop from space research are warmth-giving, spot-resistant, water-proof draperies made from electrically sensitive fabrics of a 99 per cent carbon fiber that is now used as an ablative for the heat shields in some of our rockets. Tiny half-volt transistor batteries sewn into the Space Age draperies would provide current for the heat. Such draperies would eliminate the need for the twice-yearly exercise with storm windows. The electrically sensitive material is currently in an experimental stage as far as home uses are concerned, but it has been used successfully in a chill-proof bathing suit. It also could be used for all-season children's clothing. Without the batteries, the clothing would be suitable for balmy weather. In cold weather, the insertion of batteries into the seams or pockets would provide adequate and uniform warmth.

Today's rapid advances in electronics promise wall color changes for tomorrow at the flick of a switch. Interior painting will be obsolete. Special built-in lighting will be able to transfer walls to any color, even provide patterns, without otherwise changing the colors of the furnishings or the lighting in the room.

What woman has not wished for a magic wand that she could point and have dinner cooked in the twinkling of an eye, or wave to make dirt and garbage disappear, or use to summon children who might be somewhere in the neighborhood out of reach of the sound of her voice? Such a magic wand is near at hand—another miracle of Space Age science. It is the LASER (Light Amplification by Stimulated Emission of Radiation), an intense beam of light achieved by exciting a ruby or other specific solid or gaseous material with light. It has many uses in space research, particularly in the field of communications. A one-pound gas laser that can send ten voice messages for more than a mile has been developed for possible use by astronauts for future rendezvous and docking operations in space. It would enable the space pilots to check their approach speeds continuously during this maneuver over their own private beam of light. Adapted for the homemaker, it could be used to send out calls to her family if they were within a radius of a mile. The bright beam of the laser, which is now being used to cut and weld metals and has been used successfully for knifeless surgery, may be fashioned into a kitchen-aid, a unique appliance that would cook meals, not in a minute, but in a microsecond. Another possible home use suggested for the brilliant light source is to vaporize garbage and dirt. But one would have to exercise caution in order not to vaporize other desirable matter.

Discardable clothing made of miracle fibers, small-size computers which will run households, and ultramarkets with push-button shopping are also future possibilities. Fantastic? Perhaps, but they have been predicted—not by wool-gathering dreamers but by persons of high achievement, among them Dr. Glenn T. Seaborg, and authorities such as the Department of Agriculture.

The disposable miracle fabrics may be fashioned into clothing without needle and thread. The paste-pot could ultimately replace the sewing machine since special adhesives may be used to join seams. Eventually high-fashion apparel may be produced at a cost that will enable every woman—not just a select and privileged few—to have a new costume for every occasion specially designed for her. The home computer could provide designs to suit individual requirements. After all, we now have computers that can predict weather patterns, design space ships and draw up plans for apartment buildings. It should be relatively simple to build a computer that could advise a woman on what to wear, draw patterns, and, if conventional cooking has not given way to synthetic diets, could plan menus and thus forever eliminate the worry of "What shall I make for dinner?" The home computer could also prepare the family budget, plan furniture arrangements and provide patterns for flower arrangements and wall hangings, and even suggest appropriate discipline for the children. The computer would also provide a shopping list that the housewife, by pushing a button, could transmit to the electronic ultramarket, foreseen by the Department of Agriculture, which would take and deliver the order.

It is likely that our woman of the future may even be spared the exertion of pushing buttons. Earlier this year, scientists at Johns Hopkins Applied Physics Laboratory demonstrated a very mobile, 100-pound, one-armed robot. They dubbed it The Beast and gave it a brain as well as an electrical sense of touch that enables it to replenish its energy and navigate. When its 12 silver cadmium batteries start to run down, The Beast feels its way along a corridor until it comes to an electrical outlet. With microswitch fingers on the end of its arm, The Beast senses the contours of the socket and then inserts two prongs into the outlet to recharge its batteries. When its energy has been replenished, it pulls out the prongs and moves on.

This is, admittedly, a very primitive electronic automaton, whose sole function is to keep moving and maintain itself. But its potential for the home and its application to the housekeeping burden of home-making are obvious—at least, to any woman. Its perpetual motion and endless energy need only purpose. Surely the scientists at Johns Hopkins can develop the brain of The

Beast to enable it to cope with all the burdensome household tasks—from cooking to serving, from cleaning to shopping. The Beast could be fully domesticated to take over most of the more onerous chores without fear of complaint or threat of work-stoppage—so long as electricity was freely available.

Push-button automated housekeeping is a real possibility for the future. When it comes to pass, it will free woman from the burdens of domesticity. What will this freedom mean? Will it be a dream come true or a nightmare? It will be a nightmare if the elimination of household chores induces a feeling of uselessness, of obsolescence. But this could only happen if housekeeping is mistaken (as it too often is) for homemaking.

Housekeeping and homemaking are not the same. One is a burden; the other is a challenge. Housekeeping is routine drudgery that can bring satisfaction and a degree of accomplishment only if it contributes to homemaking. Homemaking is a creative effort; and, like all creative efforts, it is an affirmation of life. In this sense it is peculiarly and exclusively woman's work. For it is woman who brings life into being. She is the homemaker —not when she cleans, but when she instills in that life an appreciation for the basic values of humanity and the desire to see those values maintained and strengthened.

Science and technology may be able to provide a machine with a brain to do housekeeping superbly and thus automate the housekeepers in our society out of existence; but they cannot devise the means to endow a machine with the heart and spirit essential to homemaking.

The freedom from housekeeping that our electronic and other technological wonders promise to give to the woman of the Space Age is her greatest opportunity and challenge. This freedom will provide the time for her to carry out her primary role as homemaker for her family, and, without stinting the emotional, spiritual and intellectual needs of those immediately dependent upon her, enable her to extend her role to the family that is all mankind.

In the eighteenth century, Samuel Johnson wrote: "Nature

has given women so much power that the law has very wisely given them little." In the twentieth century, the law has been more generous and women have gained the franchise, thus enlarging their opportunities to put their power, so long restricted to domestic service, to other constructive and productive uses. Today one out of every three American women—totaling over 24,000,000 in 1963—is employed. Approximately 60 per cent are married and 20 per cent are mothers as well as wives. These figures, incidentally, do not include those women who are employed only part time.

It was not so long ago that the attitude of society toward working wives and mothers was one of either pity or scorn. They were pitied if they worked because of economic need, and scorned as inadequate wives and mothers if they chose to work. The number of working wives has increased by more than 20 per cent since the early 1950's. The rapid acceleration in their numbers marks a significant departure from the traditional prejudice against outside employment for homemakers. It reflects an important advance for the status of women and is evidence of society's growing appreciation of their abilities.

Many of the married women who work today do so from choice, motivated by a desire for useful and interesting service rather than salary. And studies have shown that while the majority of working homemakers work for money, they also are consciously seeking employment that will provide them with a sense of achievement.

Opportunities for such employment have never been greater. There is an ever-increasing shortage of teachers, nurses, physicians, dentists, scientists, engineers, technicians, social workers and a host of other trained professionals as well as persons trained in the equally important supporting skills: clerical, business, bookkeeping, and maintenance. The need for homemakers has not lessened, but today's homemakers are proving to their own inner satisfaction and the benefit of others that they can *do* as well as merely *be* in the world around them. They are proving as adept in political science as they are in domestic science, in the fine arts as in the home arts, in social welfare as in the social graces, in office management as in home management, in the laboratory as in the kitchen. The old concept of the woman torn between home and career is disappearing. Both are now possible, and the prospects for the future—when the great gift of "released" time promised by science and technology is realized—are even greater.

This means, of course, that women will now have to prepare themselves for the creative and intellectual opportunities that lie ahead. There will be challenges, more demanding than any they have ever known in the arts, sciences, and other fields of human endeavor. But perhaps the greatest challenge that mankind will face in the future is in the art of living; and it is here that women—homemakers, if you will—can make their greatest contribution. Students of automation and our sociologists warn that as machines take over more and more of the work of men, leisure will increase. This will be a problem rather than a pleasure for the great majority of our population, even assuming that there is a satisfactory economic adjustment so that all of our citizens are adequately supported.

Those homemakers who have learned how terrible it is to be burdened with time and no purpose can now begin to explore ways to employ this coming threat of leisure so that our lives and our civilization will thereby be enriched. Constructive use of leisure will be one of our greatest needs. Unless we can enhance and appreciate the cultural and creative arts in the time that is to come, the world will be a prison. An important effort, therefore, for women today and tomorrow is to participate in those activities which some among us have labeled "frills." For the "frills" of today may well prove to be the fabric of that better tomorrow. Homemakers can and should give consideration to this future problem without curtailing their present efforts to achieve that better tomorrow.

Here in the United States where war has been declared on poverty, a corps of 4,900 homemakers are in the front lines. Associated with the National Council of Homemaker Services, they work with state and federal agencies, among the latter the Children's Bureau of the Department of Health, Education and Welfare, giving of their skills and services to families in need. Dr. Ellen Winston, U.S. Commissioner on Welfare, has saluted this homemaker service for its efforts to break the cycle of ignorance, disease, and deprivation upon which poverty feeds. But a corps is not an army; and it will take an army of homemakers to win the war—just in America.

The entire world is in sore need of homemakers, and the American woman is gaining the advantages that will help her meet that need. With the increasing abundance of time that is hers now and that lies ahead in the future, she can develop her talents and professional skills and dedicate them to the elimination of the social and political problems that have plagued mankind. If she does this with thought for the greater future, our world yet may be a home of inner and outer abundance for every man, woman and child.

FROM OUTER SPACE—ADVANCES FOR MEDICINE ON EARTH

Hubertus Strughold

The science and technology which have advanced man safely into space have brought about startling medical advances for man on earth. Out of space research have come new knowledge, techniques, and instruments which have enabled some bedridden invalids to walk, the totally deaf to hear, the voiceless to talk and, in the foreseeable future, may even make it possible for the blind to "see."

Such developments, astonishing and revolutionary as they are to us as contemporary witnesses, are but forward steps in the long time span of the history of medicine, which is almost as old as man himself. In its primitive form, medicine—the science and art of preserving and restoring health—was practiced even by the caveman. It was refined and developed into a rather sophisticated concept by the Greeks and Romans, whose philosophical speculations and anatomical observations dominated medical thought throughout the Middle Ages. Modern medicir began to take form with the invention of the microscope in tl seventeenth century. With the subsequent expansion in tl physical and chemical sciences, accompanied by such discoveri as the x-ray and anesthesia, medicine became the highly scier tific art it is today with numerous disciplines and specialties.

Modern medicine, however, is also conditioned by new an changing environments which man creates or finds and whic produce risks to life and health unknown to our ancient for bears. Environmental changes from modern industrial growt and technology have brought with them the new branches c industrial and environmental medicine. With the airplane cam aeromedicine, encompassing the medical problems associate with higher atmospheric altitudes, speed, windblast, orientatio sound stress, and other related exposures from air travel. Th development of the rocket made possible man's travel beyon the atmosphere and brought into being another branch c environmental medicine known as space or aerospace medicin

Aerospace medicine is an extension of aeromedicine along th vertical frontier. While it is concerned with the medical prob lems of atmospheric flight, it is particularly interested in prob lems which may affect the life and performance of the astronau in space flight and his ability to function and to survive o other celestial bodies such as the moon and Mars. Space medica efforts are serving not only these particular purposes, but the have provided and will continue to provide both specific an general benefits for medical practice on earth. A few of th specific by-products were noted at the opening of this chapte The general benefits include a vast increase in the knowledg and understanding of the physiological functions of the huma body in relation to the earth's environment, and improvements i medical methods, concepts, terminology in research and teachin are already evident. These benefits—both specific and general— are demonstrated in the following examination of the physiologica and medical aspects of gravity.

Gravity is the force which determines our weight and i

expressed as 1 g, the gravitational force at the earth's surface. It is so much a part of our lives that even in medical training, gravity usually has been treated as self-evident, ever-present and constant and, therefore, not a particularly exciting or significant component of our natural environment. But to explore space, man had to find ways to free himself from this environmental force; and it was not until then that he became really conscious of its pervasive influence on his health and physiological development.

Gravity becomes conspicuous in the life of man on earth only when the human organism is not yet strong enough to cope with it, as in infancy and the first two or three years of growth, or when the organism has become too weak, as in certain diseases and old age. Yet gravity is a force against which the inhabitants of earth struggle daily with every breath, and this conflict plays a very large role in human development. Indeed, man is a product of this constant, and his health and physical well-being may be measured to a great degree by his ability to adapt to the continuing demands of this force.

Modern means of transportation can change this gravitational constancy. In travel by ship on a rough sea, we are exposed to periodic variations around the geogravitational norm of 1 g; and these variations occasionally lead to seasickness. Similar g oscillations may be experienced during air travel under adverse flying conditions and may result in airsickness. Both of these types of motion sickness now are well understood and generally are therapeutically controllable. This understanding has benefited our knowledge of nausea and has advanced the development of therapeutic drugs.

In certain military flight maneuvers (turn, pull-out, ejection) increased g-values up to about 4 *g's* are experienced. In space flight increased g-loads of 7 to 10 *g's* occur for several minutes during launching and atmospheric re-entry, causing enormously increased weight. A man who weighs 150 pounds on earth will weigh 1,050 to 1,500 pounds during moments of launch and reentry. Even lifting an arm under this burden of excess weight would be very difficult. It has been necessary, therefore, to find a way for astronauts to maneuver and perform certain operating functions under severe gravity stresses. One method being explored is a program of physical training that will enable the astronaut to contract certain muscles at will. Each of these muscles, connected to electric transducers to amplify electric power, could be wired to specific control mechanisms. The contraction would supply the electrical power, a tiny discharge which would be amplified many times by the transducer, to turn the controls off and on, change the direction of the vehicle, and make it possible for the astronaut to perform many other functions which otherwise the severe stress of increased g-loads would prevent him from doing.

Research sponsored by the Veteran's Administration has been undertaken to adapt this knowledge for the severely disabled—those who are armless, for example. By training persons so handicapped to contract individual neck and shoulder muscles at will, and by amplifying the contraction with transducers wired to artificial limbs, it is believed that the handicapped would be able to manipulate the artificial limbs efficiently enough to perform manual tasks of every variety with precision.

To learn how to protect pilots and astronauts from severe gravitational stresses, experiments were undertaken on large centrifuges and rocket-powered sleds. These have proved of great value in the analysis of mechanical injuries and for protection against traffic accidents on earth. The use of safety seat belts in automobiles, which have saved many lives, resulted from knowledge gained from experiments by Colonel John Paul Stapp on rocket-propelled sleds.

Anti-*g* suits are another development from these gravity studies. They are worn by pilots of high-performance jet aircraft and space vehicles to counteract increased hydrostatic pressure (in-flight increased pressure of body fluids) and prevent blacking-out that otherwise would result from the decrease in normal blood circulation. The suits are designed to exert mechanical pressure upon the lower part of the body, especially the legs, thus maintaining normal blood pressure and circulation and preventing blackout.

In 1961, Air Force medical researchers reasoned that the anti-g pressure suits designed for the astronauts to prevent blacking-out might be adapted to restore normal blood pressure to bed-ridden victims of strokes and similar circulatory disorders, enabling them to walk and even work again. The reasoning proved correct; and today former invalids, among them house-wives, farmers, laborers, bankers and industrialists, bedridden and helpless for years, have been restored to useful life by space-suit trousers medically tailored for their health.

These are a few of the benefits, promised and realized, from studying the physiological responses to normal and increased gravity. But even more than increased gravity, sub-gravity and zero-gravity, resulting in decreased weight and weightlessness as it occurs during orbital flight in space, has really made the problem of gravity or its absence an intriguing one.

Just what is weightlessness and how is it achieved? We must keep in mind that weightlessness in space flight is not a function of the distance from the earth as one might assume. It is rather the result of a balance between earth's gravitational pull and the inertial forces of the moving vehicle and its occupants. Thus it is dynamic in character and a demonstration of the fact that mass and weight of a material body are not the same. Mass is an intrinsic property of matter representing the sum of all atomic particles that compose the material. It is, if you will, content and substance. Weight is an extrinsic property of matter depending on external forces, primarily gravitational forces. In actual space flight, weight is absent; and this is for us a strange and most exotic condition—one that our astronauts appear to enjoy. Nevertheless, human tolerance to weightlessness is a matter of interest and concern for space medical researchers.

It might be worth mentioning at this point that so far we do not find the word "weightlessness" in the medical textbooks. Yet in a free fall, in the initial phase of a descending elevator, and in a ship at the crest of the wave, weightlessness is experienced briefly. It was generally overlooked until space medicine brought it dramatically to our attention; and in the future weightlessness will be used in medical teaching as a contrast to the normal gravitational condition.

What have we learned about the effect of weightlessness on the human organism? Can this organism, so much a product of a gravity environment, adapt to the weightless, zero-gravity environment of outer space?

Those who have experienced space flight to-date, ranging in time from hours to several days, have demonstrated that they could perform efficiently in a weightless condition. None of those who have orbited the earth were conscious of any adverse physical reaction to weightlessness, with the sole exception of Cosmonaut Ghermain Titov who became dizzy and nauseous after three hours in orbit. To counteract the nausea, Titov was advised to keep his head still during the flight; for, significantly, in space it is one's own motion that causes the sickness. He fully recovered only when he re-entered the earth's atmosphere and experienced the force of gravity. It would appear, therefore, that tolerance to space flight, like tolerance to ocean or air travel, may vary from individual to individual. Titov's experience, however, has provided us with a better understanding of the otolith organ, located in the inner ear, which affects physical balance and is gravity-dependent. The otolith organ is one of our mechanoreceptors, which are organs responsive to gravity. Other mechanoreceptors, such as the touch sense of the skin and the muscle sense, are not gravity-dependent. This explains why the manual performance of the astronauts was not impaired by weightlessness.

While none of our space travelers except Titov were conscious of any adverse reaction to weightlessness, there have been significant developments of medical concern. For example, the urge for voiding is vastly reduced in a zero-g environment, and prolonged delay in voiding can result in damage to health. However, it appears that fluid elimination can be induced when necessary if a conscious effort is made even in the absence of the stimuating effect of gravity. Whether this would hold true for long-erm space voyages remains to be seen; but judging from the imited experiences to-date, this effect of weightlessness may not be a significant barrier to manned space flight. It does explain, however, why after surgical operations patients should be brought to a vertical position as soon as possible to take advantage of the stimulating effect of gravity to induce voiding.

When exposed again to the 1 g of earth after their flights in pace, some of the Mercury astronauts and the Soviet cosmonauts experienced a loss in bone calcium, muscular weakness, and a pooling of blood in the veins of their legs—symptoms of what s termed orthostatic intolerance. Some of these symptoms lasted or hours and even, in some instances, days; but they eventually disappeared. Similar effects have been observed in healthy vol-unteers subjected to experiments of prolonged bed-rest as well as n patients who have been confined to bed for long periods of ime. Muscles and bones atrophy when not used; and it may be hat inactivity is the main cause of calcium loss and muscular weakness experienced by those exposed to space flight. Weight-essness may be only a contributing factor. It is, however, con-ceivable that without preventive measures, those exposed to long periods in space might suffer serious decalcification. To prevent uch effects special exercises have been developed. Because of he limited confines of a space vehicle they are of the isometric ype, found to be very useful for bedridden patients who cannot move about and perform exercises in the isotonic fashion. Iso-onic exercises are walking, running and dancing, which involve changes in the length of muscles but not too great a change in heir firmness or tone. Isometric exercises include fist-flexing, pressing against a wall or some other counteraction against elastic force. Such muscular action results in increased tone.

For another example of the possible effects of weightlessness, et us assume the birth of a baby of terrestrial inheritance either under zero-g conditions of a space station or under the sub-gravity condition of the moon, which is one-sixth that of earth. Brought to earth after several years on the moon, the subgravity moon baby probably would lack the necessary firmness of muscle and bone to walk against the 1 g on earth. It is likely that he would first have to learn to swim and later to walk, following in a sense the developmental pattern in the evolution from sea life to land life. A similar course in the development of movement can be observed in polio-inflicted babies.

It also might be interesting to speculate on what the physical structure of man and his behavior would be if earth had a gravity such as that of Mars which is .38 g, or Jupiter, which is 2.64 g. Experiments have demonstrated that mice, rats, or chicks born in an environment in which the gravity has been artificially increased above the norm do not grow as large as those born in a normal earth-gravity environment. If the converse should prove true—that animals born in a lower-gravity environment would grow larger—we might expect organisms born on Mars to be three times larger than the same organisms born on earth, while on Jupiter they would be less than half the size of earth-born organisms. These conjectures are, of course, subject to some qualifications, since a change in gravity also would influence other physical environmental factors vital to man, such as the atmosphere, its pressure and chemical composition.

Medical research for manned space flight also has brought a better understanding of the fascinating problem of the physiological day-night cycle. After a number of hours of activity, man requires rest and sleep for the restoration of energy; and this rest and activity cycle usually is synchronized with the physical day-night cycle. Keeping this cycle functioning is a physiological law, and we even speak of a "physiological clock." But neither in space nor in the deep sea is there a sequence of day and night. In the deep sea, below 500 meters, there is eternal night. In space, beyond the atmosphere, day and night occur, so to speak, at the same time: a permanent bright sun shines from a permanent black sky. Only when the astronaut moves through the shadow of the earth is there a short exclusive night of about thirty minutes.

The astronaut still needs sleep and rest after his strenuous activities. His rest and activity cycle would probably work into a "free running cycle," still within the temporal frame of twenty-four hours with seven hours of sleep, but not necessarily synchronized with the day-night cycle observed in any particular time zone on earth. The same would be true on the moon, where the day-night cycle lasts for about twenty-seven days. Thus, since the space traveler is living beyond our geographical time zones, the study of the physiological nature of man's day-night cycle on earth—of his physiological clock—has become important.

The day-night cycle is even important in today's intercontinental travel by air. After a transoceanic flight, that is, after crossing five or more time zones, a traveler experiences a phase shift between the geographic day-night cycle of the new location and his physiological cycle. He then is in a state of what can be called asynchronosis. It takes from four to six days to become completely adapted to the new time zone, or to become resynchronized. This is a new point in travel medicine and has some interesting consequences.

In order to function mentally and physically at his maximum efficiency, a diplomat or businessman should leave two or three days in advance of an important conference to be held on another continent so that he may in that period become adapted to the local time. Or he should pre-adapt himself by retiring every evening, several days in advance of the trip, according to the time on the other continent, so that he will not be handicapped by a day-night cycle asynchronosis when he meets with his synchronized counterpart of the local time zone. If these measures cannot be taken, the traveler should keep in mind that after eastbound flights the afternoon hours, and after westbound flights the morning hours, are the best times for scheduling important meetings for the first few days. Who knows but what East-West agreements may have been hampered because of the effects of asynchronosis on one or the other of the negotiating parties? It has been reported that actors, athletes, and even race horses are not at their best when in the state of incomplete cycle adaptation. Statesmen also are not immune to their physiological clocks.

These experiences and observations in space and atmospheric flight have led to intensive studies in laboratories and caves in which man and animals have been exposed to different lengths of artificial day-night cycles and phase shifts and the various body reactions such as temperature, metabolic rate, circulation, etc., have been observed. Even isolated tissue cells show a day-night cycle as manifested in a different multiplication rate, with one exception—cancer cells. These, it has been found, always multiply at a maximum level and no longer obey geobiological laws. This gives the medical researcher still another insight into the nature of cancer cells; and if further studies should determine that element or substance in cells which responds to day-night cycles, it may be possible to develop new methods for the treatment and control of cancer.

The rest and activity cycle is also a subject of growing importance for normal daily living and for the care and rehabilitation of the sick. Patients who have undergone major surgery notice that for at least a week, due to pain-relieving drugs, their rest and activity cycle is "out of order." They also may notice that as soon as their sleep is again confined to night time, they have the feeling that they are on the road to recovery. The recognition that a certain amount or "dose" of sleep, appropriately placed in the day-night pattern, is essential to well-being already has led to the exploration of artificial methods to induce sleep electronically, by new drugs which have no side effects, or by hypnosis. Russian experiments have demonstrated that electronically induced sleep is so deep and restful that the need for sleep can be vastly foreshortened. Properly applied and used by gifted persons, this reduction in sleep requirements could mean a greater outpouring of creativity. Drug-induced sleep has definite therapeutic value, particularly for the ailing. Sleep needs also can be reduced by hypnosis; but considerable study must be undertaken to understand the full consequences of this phenomenon.

The visual panorama in space with its black sky and perpetually bright sun is of special interest to ophthalmology. In the visible section of the solar radiation spectrum, the human eye is an indispensable and unsurpassable sensor in the exploration of space. But there also is a hazard. Looking into the sun can produce a retinal burn similar to that which occurs from looking at atomic flashes or a solar eclipse with an insufficiently smoked glass. In some cases such exposures produce a blinding effect which can last for minutes, or even cause permanent visual impairment. For this reason, astronauts reported that they always avoided looking directly into the sun. The potential hazards necessitate protective measures, preferably in the form of automatically functioning, light-absorbing, photoreactive glasses which change their transparency quickly according to the level of illumination. The development of such protective glasses now under study will undoubtedly influence the manufacture of sun glasses. Incidentally, studies of glare and blinding effects also may prove beneficial for safety in automobile traffic at night.

As in so many cases, that which is hazardous to man also can be made beneficial to him. Simulated artificial solar rays concentrated into a minute intensive beam are now used in ophthalmological clinics to fixate detached retinas by heat coagulation of the surrounding retinal tissue. They are also used to destroy small retinal tumors, thus making it possible to avoid eye surgery. Very recently a relatively new method to produce very powerful rays by LASER (Light Amplification by Stimulated Emission of Radiation) has been used successfully for the same purpose.

The technique of instrumentation for space flight is also worthy of note. Because of the original limits of our rocket power, space payloads had to be restricted in size and weight. Our engineers expended great effort in miniaturizing rocket and satellite operating components and instruments while, at the same time, striving for increased efficiency. Minuscule batteries were developed to operate satellite instruments, and these were quickly adapted for medical use in space and on earth.

For space travel, medical researchers have developed miniature instruments to record environmental factors within the capsule, such as oxygen and carbon dioxide pressures, humidity, and temperature. These have aided in developing life systems to keep the total environment within physiologically acceptable ranges. Oxygen sensors developed specifically for aerospace medical purposes now are widely used to check the respiratory air of patients under anesthesia. From these we have also gained a better knowledge of the maximum permissible level of oxygen pressure, and we know at what levels it may be tolerated for a specified number of days, without becoming toxic. We have also discovered that prolonged use beyond several days at levels previously believed to be acceptable may cause blood damage, and that carbon dioxide, which is exhaled in respiration, must not rise above a determined level or suffocation will occur. Both of these discoveries are important for patients under an oxygen tent.

The recent development of the hyperbaric or high-pressure chambers for heart surgery is another direct outgrowth of aerospace medicine. Introduced first in Holland, these chambers, which have been used with such dramatic success in life-saving heart operations on blue babies, resulted from pioneer air-pressure and oxygen-pressure studies concerned with exposures anticipated in high-performance aircraft as well as spacecraft.

Miniaturization for space also has made possible a system of medical electronics for the continual and simultaneous monitoring of all the various physiological processes—body temperature, heart activity, brain activity, pulse, blood pressure, and others—transmitting and recording data to one central place. Adapted for earthly use, such a system is being used to monitor handicapped workers on the job for energy consumption and other vital bodily responses. This type of mobile Space-Age "living laboratory" is providing researchers with new data on the physiological and psychological traits of both disabled and non-disabled workers. Now being adapted for future use in hospitals, such medical electronics may advance the diagnostic process and make it possible to sense critical conditions in advance and to signal warnings, thus expanding the potential of preventive medicine.

Biocoustic experts, specialists in hearing, found that the miniature batteries developed for space could be used to vastly improve hearing aids. Further experiments have demonstrated their benefit to the totally deaf, for surgical implants of these tiny power plants have brought sound to the soundless. Also used to power artificial larynxes, these minute batteries—some smaller than your thumbnail—have enabled the previously speechless to speak and are helping tired hearts to beat with renewed vigor.

A miniature valve developed for missiles now is being used to replace a defective valve in the human heart. Another medical aid is a viscometer, a new tool for studying blood, developed from the guidance system of the Polaris missile. It has revealed important new facts about the flow of blood vital to understanding the mechanics of blood circulation involved in heart disease and other circulatory disorders.

Another bonus from the miniaturization imposed by our rocket power limitations has been reported. This is the development of spectacles for use as "seeing" aids to the blind. They are made with tiny photo-cells linked to minute batteries which amplify differentials in frequency and intensity of sound reaching the ears of the blind person as he moves toward or away from particular objects. Thus it is possible for him to "hear" where he is going. With improvements, such spectacles conceivably would enable a blind person to walk alone without bumping into objects unaided by either a cane or a "seeing-eye" dog. The eyes of one test subject were sealed over with thin sheets of lead so that there was absolutely no question of vision. He then put on the spectacles and was able to locate a three-inch by three-inch square of white paper on an oak table—by candlelight.

Theoretically, it is also possible that by means of highly sensitive electronic devices—an electric eye if you will—impulses can be driven to the optic nerves thus bringing true sight to the sightless. To accomplish this, however, a way must be found to establish contact between the optical nerves and the mechanical equipment. It is a difficult problem, but with the continuing advances in electronics expected from space research, it is not outside the realm of possible solution for the future. As noted earlier, the pressure sense of the skin—the sense of touch—plays a vital role in space operations. Research on the sense of touch indicates that more efficient utilization of this pressure sense, enhanced with the aid of our new electronic devices, may enable the blind to "see" with their fingers and even distinguish colors.

The application of electronics to medicine, now in its beginning stages, is influencing new concepts and theories concerning physiological functions themselves. This new field is called bionics: life functions interpreted from the standpoint of electronic theories which someday can be adapted to extend our senses.

For the purpose of logistics in space cabins, methods have been developed to regain the oxygen from the exhaled carbon dioxide and water vapor. For short durations, that is, for a period of weeks, physiochemical means are employed to recycle the respiratory gases. In extended space operations, lasting many months, especially on extra-terrestrial bases such as a lunar base, regeneration of respiratory gases by biological methods similar to those we observe in nature becomes a necessity. Here the process of photosynthesis found in all green plants enters the picture of medical research, and we are learning to reproduce the properties of the earth's "macroclimate" on a "microclimatic" scale which might be useful in many ways, for instance, in disaster shelters.

The final goal of astronautics is, according to the meaning of the word, a flight to another celestial body. This has raised the problem of interplanetary contamination—contamination of the earth via rockets by microorganisms from other celestial bodies and vice versa. Methods are being developed and employed for the sterilization of rockets, and such sterilization on a large scale with new chemicals may bring some new ideas for medical sterilization. This area of research is the link between space medicine and another important scientific field in the Space Age: astrobiology or exobiology, the study of indigenous life on other planets. Such studies must include the past—paleoastrobiology. For these the earth's atmosphere presents an interesting model for the conditions on a life-supporting planet from its primeval hydrogen phase to the present oxygen phase; and it offers a stimulating platform for general medico-biological considerations. Bacteriology provides an interesting example.

In all probability there was a time during the Proterozoic era some two billion years ago when the carbon dioxide content in the atmosphere was considerably higher than it is today. Recent bacterial studies have revealed that an increased concentration of carbon dioxide generally promotes the growth of bacteria. For many bacteria the optimum lies between five and ten percent of carbon dioxide by volume. The pneumococcus belongs in this category. It is, therefore, not surprising that this coccus finds an ideal environment in the lungs for a population explosion, producing a severe pneumonia within a few days, since the alveolar air of the lung is much richer in carbon dioxide than the ambient air. The carbon dioxide optimum of bacillus tuberculosis lies between two and three percent by volume. Carbon-dioxide-philic bacteria are probably very old paleontologically. If so, when inhaled, they return to their original medium, preserved in the inner atmosphere of our lungs or in certain tissues. Paleobiological and astrobiological studies will undoubtedly extend our knowledge of terrestrial bacteriology and aid in eliminating certain diseases.

Science is more than an abstract search for knowledge and truth. Its higher purpose is to serve mankind. This is particularly true of the science of medicine. Research in space medicine is helping unravel the marvelous mystery that is the physical man and in so doing is making its contribution to the realization of manned space flight. But while this is its proper immediate mission, the higher objective is not merely man's survival in his journey to other worlds; it is to assure that he will have at his command for this bold venture the full resources of both mind and body. For man in space as it is for man on earth, the goal of medicine is to achieve fulfillment of the ancient Roman proverb: *Ut sit mens sana in corpore sano.* Let there be a sound mind in a sound body. Through space medicine we are coming closer to this incorporation that will bring full health to man and may give him the power to perfect his way of life on earth.

A LOOK AT THE WEATHER FROM OUTER SPACE

S. Fred Singer

Weather is indisputably the sovereign physical force on earth. Everyone—rich or poor, high or low—is affected and, indeed, dominated by the meteorological phenomena we call weather. Its influence is ever-present and all-pervasive. It has upset the outcome of wars and political elections, as well as horse races. It has caused the postponement of social events and satellite launchings. Even the strength and vitality of nations are influenced by its disruptive effects on life and property.

Mastery of weather—its prediction and control—has long been a dream of men and nations, and the weather satellites developed by the United States are providing observations of the atmosphere that one day may make that dream a reality. These orbiting weather observatories have given our nation a preeminence in space exploration which, in my opinion, far overshadows Soviet achievements in manned space flight. For the nation that will be first to predict and control weather need never fear being second in influence and prestige. Its primary position among the family of nations will be assured.

Weather phenomena, whether they be rain, sleet, snow, balmy winds, drought, or hurricane, are caused by the movement and interaction of masses of air in the atmosphere. These movements have been described as a never-ending series of battles between warring masses of air: high-pressure and low-pressure masses, cold fronts and warm fronts. Weather is the result of these battles. To predict the outcome of these forays and thus forecast weather, it is necessary to trace the movements of air masses, to measure their temperature and pressure on a global basis, and to collect and analyze the observed data rapidly.

Man's efforts to master weather are part of his ancient history. Primitive societies had their rainmakers. The Pharaohs of ancient Egypt worshiped the Sun God. The first scientific study of weather was made in the fourth century B.C. by Aristotle. This remained the authoritative work on weather until the invention of the thermometer by Galileo in the sixteenth century, and the barometer in the seventeenth century by Torricelli, which made possible measurements of temperature and air pressure.

In the early days of our nation, Thomas Jefferson and Ben Franklin exchanged weather data by mail, carried by horse-drawn conveyances or by foot; the information was obsolete by the time it was received. The invention of the telegraph in 1844 was the first step toward modern forecasting, for it made possible the rapid gathering and transmission of information from many areas and led to the establishment of a weather-reporting network. Congress, recognizing the importance of weather, established in 1870 a national weather service which in 1890 was officially titled the United States Weather Bureau.

Collecting of weather data was, at first, limited to observations made from the earth's surface. Later, kites, balloons and aircraft were used to take measurements higher in the atmosphere in various parts of the world. More recently, radiosondes, radar, and meteorological rockets made measurements possible at even higher altitudes, while computers speeded the transmission of data and its analysis. The satellite, the latest device for weather data gathering, has already proved a valuable complement to the more conventional techniques of weather observation.

The conventional weather observation network over the earth has a spacing that ranges from a few miles in densely populated Western Europe to a few thousand miles between ships and islands in the oceans. In the Southern Hemisphere, where there are very few stations, there are large gaps in data collection. Weather prediction has been hampered because the ocean areas, which cover some 70 percent of our planet and which are the breeding ground of much of our violent weather, are largely unobserved. The hurricanes which strike our eastern seaboard and the Gulf Coast originate over the Atlantic Ocean.

Weather over the oceans can be observed from surface weather ships, but a comprehensive system of such ships and their constant maintenance at sea is extremely expensive. A weather ship which costs one million dollars to maintain and operate annually can observe an area with a maximum radius of about 30 miles. A weather satellite which costs approximately four and a half million, including its launch into orbit, can observe an area of approximately 640,000 square miles, collecting information over vast expanses of ocean and remote areas of land where no other means of observation are possible. To date (October, 1964), the United States has successfully launched eight TIROS (Television and Infrared Observation Satellites), and is now embarking on an operational satellite system based on a modification of TIROS.

The more rapid discovery of hurricanes by TIROS satellites makes possible advance warnings that permit greater protection for lives and property. During the summer of 1961, TIROS III discovered Hurricane Esther and tracked it and seventeen other hurricanes, typhoons, and tropical storms. Information from TIROS about Hurricane Carla in September 1961, combined with conventional weather observations, made possible the largest mass evacuation in the United States. This evacuation from the coastal sections of Louisiana and Texas was a major factor in the resulting very low death toll from a storm which caused close to half a billion dollars in property damage. TIROS V and VI were used routinely and successfully to hunt for and track tropical storms during the summer of 1962.

The successful orbiting and operation of the first TIROS on April 1, 1960, was the culmination of more than ten years of speculation, planning, and experimentation by American scientists. In 1947, pictures of large cloud systems in photographs recovered from cameras flown in V-2 rocket nose cones made meteorologists aware of their possibilities for weather forecasting. They recognized, however, that rockets were not practical for getting such pictures. In 1951, scientists of the Rand Corporation reported that pictures of clouds taken by a satellite and relayed to earth by television would be economically possible.

Design of the TIROS was initiated by the Advanced Research Projects Agency (ARPA) of the Department of Defense in May of 1958. In October the TIROS project was taken over by the newly established National Aeronautics and Space Administration, as were many other basic scientific space projects. Seventeen months later, TIROS I, the first fully equipped meteorological satellite, was launched successfully and within two hours sent its first television pictures back to earth.

The present TIROS is shaped like a giant pillbox, 42 inches in diameter, 19 inches high, and weighs 285 pounds. Covering the top and sides of TIROS are 9,200 solar cells which transform sunlight into electricity to recharge the batteries which power the satellite's electronic systems. Each TIROS carries two cameras. Some have radiometers to measure reflected solar radiation and terrestrial infrared radiation.

Each TIROS orbits the earth once in 100 minutes at an average altitude of 450 miles. The satellite is stabilized in space by spinning on its vertical axis 12 times per minute. Thus the cameras, which are mounted parallel to the spin-axis, face toward the same point in space during each orbit. This means that the cameras can "see" the earth roughly 25 out of each 100 minutes, and that the pictures are not taken straight downward, but at some angle from the vertical.

The television cameras take a series of overlapping pictures, each of a square area ranging from about 800 miles or more on a side. The television cameras of TIROS take still pictures. When the shutter snaps, in 15/10,000 of a second, an image is formed on the face of a vidicon tube. This image is scanned across and from top to bottom by an electronic beam; the scanning changes the visible image to electrical signals which are recorded on magnetic tape right in the satellite. The entire picture is made up of 500 of these scan lines, which the scanner "reads" from the vidicon tube in two seconds.

Operation of the satellite consists basically of one command to the satellite to take pictures, either in a remote area or over a ground station, and a second command to transmit the pictures and other data to the ground. The decision as to what region is of greatest interest on each orbital pass is made by the U.S. Weather Bureau. The TIROS Technical Control Center, under NASA's Goddard Space Flight Center, "programs" or computes the command signal to be sent to the satellite. This information is passed to the Command and Data Acquisition (CDA) Station at Wallops Island, Virginia or Fairbanks, Alaska. The CDA Station transmits the signals to the TIROS satellite as it passes overhead. When the satellite reaches the proper area, the clock system in the satellite triggers the camera. Every thirty seconds after start time the camera takes a picture until 32 pictures per camera have been recorded on the magnetic tape on board. A limit of 32 pictures per camera is imposed by the capacity of the magnetic-tape storage. As the satellite returns to within telemetry range, the acquisition station commands the satellite to transmit stored pictures and other data to the ground, to take and transmit direct pictures if the station is in a daylight area, and finally, sets the clock for the next orbit.

On the ground the pictures are simultaneously recorded on magnetic tape and displayed on a kinescope (TV-type tube). The kinescope is photographed and the pictures are recorded on 35-mm film. Within fifteen minutes after receipt the film is developed and prints are made. These prints must then be fitted to geographic latitude-longitude grids which are prepared on electronic computers at the readout stations. These grids are prepared prior to the reception of the pictures on the basis of predicted orbital data. The pictures are placed on top of the grids on a light table, and then they are drawn on each photograph in ink. In this way, the gridded pictures can be fitted together as a mosaic from which a nephanalysis or cloud map is then constructed. Elements of the cloud picture are transposed by hand to produce the cloud map on a usable meteorological scale.

The cloud map is sent from the readout station by facsimile to the National Weather Satellite Center near Washington, D.C., where it is then incorporated into regular analyses of the National Meteorological Center. From here the cloud map is retransmitted by land-line facsimile to weather stations all over the United States. The cloud maps are also broadcast routinely by radio facsimile toward a number of areas, including Europe, North Africa, the Pacific Islands, Eastern Asia, Australia, and South America. They are coded and transmitted internationally by teletype and radio to those stations which are not equipped with facsimile receivers. All of these Weather Bureau transmissions start within two and one-half hours after the pictures are sent by the satellite to the ground stations. This guarantees that the information will be timely.

The pictures and the maps made from them show the meteorologist the large-scale cloud systems over the earth. These cloud systems are remarkably persistent, showing slow changes in shape from day to day. Knowing where these cloud patterns are, how they are moving, and how the shape is changing enables the forecaster to predict where they will be and what effect they will have on tomorrow's weather.

The pictures from TIROS gave weathermen their first actual view of large-scale cloud patterns. Previously the organization of the cloud patterns associated with various weather phenomena had been deduced from ground observations and theoretical considerations. It was indeed gratifying to find that the cloud pictures in general confirmed these deductions. However, some of the details of the cloud patterns were quite different. The TIROS pictures also revealed totally unexpected small-scale cloud arrangements, such as "square" clouds, which are indicative of active thunderstorms and possible tornadoes. Beautiful ripple patterns are seen on the lee side of mountains; these are associated with turbulence, extremely dangerous to aircraft flying in these areas. Seen also are feathery, swirling vortices, crescent and doughnut-shaped clouds, radial clouds shaped something like a many-armed sea spider, and many other patterns never before seen. These are all the subject of research; some have been explained, some are still a puzzle.

The very existence of clouds usually implies that there are upward vertical currents of air in their vicinity. Clear streaks in spiral cloud patterns suggest downward motion and drying out of the air. Many other dynamic motions can be estimated merely by looking at the cloud pictures. Research meteorologists are discovering heretofore unexpected details about the circulation of the atmosphere through the study of these pictures. The organization of tropical storms in their developmental stages is slowly emerging. The circulation in the tropics, for example, is proving much more complex than previously thought.

Radiation data from TIROS are also being studied and compared with the measured energies that drive the earth's atmosphere. They should lead to a better understanding of how the sun's heat is used by the atmosphere and the development of forecasting techniques that some day may permit us to predict weather a season or even as much as a year in advance.

The list of things that meteorologists would like to do with the data available from the weather satellites is growing every day. The actual use of the dynamic motions revealed in cloud pictures awaits development of methods to derive quantitative values for them to replace the qualitative estimates we now can make. The main limiting factor in solving this and other problems is the time and money available for research. The other limiting factor is our incomplete knowledge of all the mechanisms that cause our weather. An intensified research effort will lead to the solution not only of known problems but also may result in a breakthrough in areas now unknown.

One such area is weather modification. With present knowledge, we can under ideal conditions produce rain or dissipate clouds. Rain can be produced by "seeding" clouds with dry ice or silver iodide smoke. Dry ice freezes some raindrops in the cloud, which in turn freeze others, thus starting a chain reaction. The ice crystals formed attract liquid cloud droplets. Finally, the ice crystals and the droplets they collect become large enough to fall out of the clouds as rain. The particles in the silver iodide smoke are crystalline in form. When the smoke goes into a cloud, ice crystals form on these particles and raindrops build up just as when dry ice is used.

Clouds can also be dissipated by these materials when the cloud temperatures are very cold. In this case the ice crystals induced by these materials fall out of the cloud as a very fine snow, which may evaporate before hitting the ground. The production and fall of the ice crystals is extremely rapid and the cloud literally falls away.

Weather control in general, however, is not understood. Although we know that a single thunderstorm generates as much energy as one hundred atomic bombs of megaton size, we do not know how to control such forces, either by chemical or physical reactions. The forces in the atmosphere are so closely linked that a simple modification of weather in one small area could conceivably unleash a destructive reaction half a hemisphere away. If we could bring rain to the Mojave Desert, for example, the disturbance it might create in the great planetary wind belts might cause a drought in Illinois, tornadoes over New England, destructive winds in France, dust storms in the Urals, and so on. Only new basic knowledge will tell us whether it will be possible to modify weather at will without courting disaster—whether it will be possible to bring a desert to flower, to break up a hurricane or tornado, or to moderate severe winters and unbearably hot summers. Satellites someday may provide that knowledge.

TIROS television cameras have given us pictures not only of clouds but of ice in waterways, snow, smoke from forest fires, and, of course, land areas. Pictures of dry land areas, such as the Mediterranean Coast of North Africa, the Arabian Peninsula, and Western Australia have shown unexpected detail. The oases, varicolored soils, and sharp coastal outlines which are revealed can be used by map makers to fill in details in unexplored areas.

Pictures of the Gulf of St. Lawrence taken by TIROS I showed ice floes, pack ice, and cracks and leads in the ice fields. These pictures have indicated that satellites could be used to do ice surveys, and even to predict changes in the ice fields. Detailed study of the pictures, to some extent, makes it possible to forecast the breakup of the ice, since the intensity of reflection from the ice changes just along the line where the ice breaks off. We have had a joint project in operation with the Canadians since the spring of 1960, and estimates are that this type of project can save about $1.7 million annually through more efficient utilization of ice-breakers and reconnaissance aircraft. If photographs from satellites show that the ice is not breaking up, there obviously is no point in sending reconnaissance aircraft or icebreakers. If the ice already is cleared out, again there is no point in sending them. The satellite can spot the intermediate regions that need either further surveillance or suggest the use of icebreakers.

Snow, as well as ice, can be seen in the pictures. Snow fields on mountain ranges have very distinctive patterns. Free from snow or shadowed by the neighboring heights, the valleys appear dark against the bright snow-capped ridges. Of course, it is sometimes difficult to distinguish between snow and the equally bright clouds. However, successive photographs of the area can be compared, since the cloud coverage changes from day to day and from hour to hour while the snow generally does not. By studying these pictures, experts can calculate potential run-off. This is quite important, particularly to our western states where large amounts of water are required for irrigation.

The cost of water in California is $15 per acre foot; the amount of water stored for irrigational purposes in all of the western states is 107 million acre feet, an amount worth roughly $1.5 billion per year. More efficient use of run-off could result in great monetary savings. The manager of a dam in the Los Angeles area, for example, has to decide by March or April whether to let water run out of the dam, depending on how much spring runoff is expected. If water is let out and the expected runoff does not come from the mountains, then water has been wasted. If water is not released and excessive runoff does occur, then there is a flood and much damage.

Prediction of runoff is made by estimating the depth of the snow. At the present time the California Water Department sends large crews of men into the mountains to take snow samples by sinking long pipes into the ground at numerous locations, weighing the amount of snow in the pipe and estimating the runoff from this measure. This is a time-consuming and expensive process. Expert estimates based on observations from weather satellites indicate that they will provide an easier, less expensive and more accurate method for predicting runoff.

Weather satellites can also provide information that will take the guesswork out of choosing the best time to plant crops. Proper planting time is crucial in certain parts of the world such as India, where the planting season is dictated by the onset of the monsoon. If the farmer plants too soon before the rainy season, the seeds will dry up. Planting too late means an inferior crop because plants will be deprived of the full amount of rain during the growing season. Data from weather satellites can be used to indicate the optimum planting season.

Weather satellites can also be used to detect the approach of locusts which still threaten great areas of Africa and Asia in swarms so thick that they cloud the sun. Locust clouds are a semimeteorological phenomenon. The insects are carried by the wind, and satellites can provide data that make it possible to predict winds in equatorial zones and thus predict the path of the locust clouds in countries of Africa and Asia which have no weather stations to speak of. Crop losses from locusts in these countries which can least afford them are extremely high. The annual cost of spotting and control is about $15 million, which yields little benefit because of the difficulty in communication and identifying the swarms. Satellites can provide accurate and rapid warning, thus enabling swift employment of methods to combat the locusts.

Another potential application of weather satellites of great economic value is forest fire detection. The annual loss from forest fires in the United States is between $50 million and $300 million; the annual cost of fire control is $140 million. The technique now used to spot fires is by means of lookout towers in forest areas. Since approximately 90 percent of the damage from forest fires is caused by the three percent of the fires which get out of control, it is very important to spot a fire early. Future TIROS satellites may spot fires by means of microwave equipment which will detect the high temperatures of fires through the smoke and steam clouds that precede the full outbreak of flames.

The value of the data received from TIROS satellites exceeded all expectations. The high performance of each TIROS encouraged the conversion from an experimental to an operational system, and the two-TIROS system was initiated. This was achieved by phasing the launching of two TIROS satellites so that essentially continuous cover of the Northern Hemisphere was obtained. While this is only 50 percent operational in terms of total global coverage, it is a major step forward. This system is the forerunner of the TIROS Operational Satellite (TOS) system which will use a satellite similar to the TIROS. The satellite will be placed in a polar orbit and will provide photo coverage of the *entire* earth. The TOS satellite will, in essence, be a TIROS turned on its side with its spin axis at right angles to the orbital plane. In effect, it will roll along the orbit on its side, hence the name "cartwheel" TIROS. The cameras will be mounted to point outward from the rim. Once each time the satellite turns on its axis, each camera will point straight down for picture taking. The cartwheel satellite will be able to provide the same coverage planned for the Nimbus satellite, a research satellite being developed to test new or improved meteorological sensors prior to their use in operational satellites.

The orbital path of the TOS and Nimbus satellites will be inclined 80 degrees to the equator. This will allow the satellite to keep its orbital plane fixed with respect to the sun line, so that it will pass over any given point on earth at the same time each day; for example, it could cross the equator at local noon or local midnight—or 3 A.M. and 3 P.M. The satellite data will be acquired by a readout station in Alaska or eastern United States and then transmitted by a rather wide-band data link to a processing center.

The Nimbus satellite will test improved television and infrared systems. It will carry three cameras whose axes will always be oriented along the satellite path. The cameras will always point earthward so that they can provide improved photo coverage of the entire earth. The infrared equipment will include an elaborate infrared spectrometer now being constructed which will measure emissions from carbon dioxide bands at several different levels or wavelengths in the atmosphere. Each wavelength has a different absorption coefficient, thus providing energy measurements at different levels in the atmosphere and revealing the temperature at each of these levels.

An ideal weather satellite system must have the capability to locate storms in all stages of development and to pass the information quickly to the ground. One difficulty is that storms not only vary in size but in duration. A tornado, for example, is a very short-lived storm which moves rapidly. You have to look at it very frequently in order to forecast its path. Thunderstorms are medium systems which last a few hours. Hurricanes are larger systems which must be checked every day. Finally, we have the extratropical cyclones, or storms of the midlatitudes, which are very large, slow-moving systems. These can be checked every few days, since they are quite persistent. Thus, the present satellites, which look at the situation about once a day, can cover large cyclones very well. They can see thunderstorm cells but cannot really follow them as they move. In order to do this, we have to be able to look at the meteorological situation more frequently. With two operational satellites in orbit, every part of the earth will be under observation twice a day; with three, each area will be viewed three times a day. More frequent daily meteorological observations can be made by a satellite in synchronous orbit at an attitude of 22,300 miles. There it will move with the same angular velocity as the earth; it will, in effect, stay "fixed" over a specific point at the Equator.

United States weather satellite research has benefited not only our own country but most of the world, including the Iron Curtain countries. We have been sharing our TIROS weather data with all countries through the World Meteorological Organization, and weather advisories and other information have been transmitted to the weather services of other countries lying in the paths of hurricanes, typhoons, and tropical storms. This has, of course, been invaluable in fostering international understanding and cooperation; and as our weather satellites improve in capability, we will be able to provide even better weather data to the nations of the world, particularly those newly emerging states that do not have adequate weather or communications systems.

It is becoming quite clear that the development of an operational system of weather satellites will vastly expand weather prediction. Even if limited to accurate forecasts for only five days, such foreknowledge would mean a total annual cost savings of billions of dollars for agriculture, the lumber industry, land and water transportation, retail industries, and water resources. The future in weather research is as unlimited as space itself; and based on present advances it holds great promise.

In his development of weather satellites man has come a long way from attempts to control weather by appealing to primitive gods. But the dream of mastering weather, though nearer reality, is still beyond reach. Perhaps the control of nature we seek by mastering weather may elude man until he achieves self-mastery. Both goals are worth striving for.

INTERNATIONAL SATELLITE COMMUNICATIONS

Thompson H. Mitchell

Mankind's progress can be measured by several yardsticks, not the least of which is communications. Communications constitute the exchange of thoughts and ideas by which men and nations have sought to achieve a community of understanding essential to harmonious relationships and to human survival.

The effectiveness of communications depends to a large degree upon the nature and the efficiency of the technical and operating facilities for transmission. An idea, however significant or useful, is of little value in a vacuum. It must be conveyed beyond its original source to have an effect, and transmission facilities provide the passageway, the road, if you will, for this purpose. Recent scientific developments in the communications satellites have made available a new passageway of vast dimensions. These satellites offer an unprecedented potential for the expansion and improvement of communications throughout the world.

In the Space Age, with its emphasis on international affairs, rapid and efficient international communications are vital to the conduct of governmental, commercial, and personal affairs. The "hot line" circuit between Washington and Moscow, which is provided via radio facilities by RCA Communications, Inc., is an excellent example, but it is only an indication of the close network of international communications that will be possible with the introduction of communications satellites. The development of a commercial communications satellite system is now in its infancy. Yet it marks the beginning of an era in communications which already has seen the transmission, between ground terminals thousands of miles apart, of live television and of data at a speed of 1,460,000 words per minute.

The launching of Telstar I on July 10, 1962, by the National Aeronautics and Space Administration for the American Telephone and Telegraph Company was a dramatic new chapter in Space Age communications. Telstar I was a low-level active repeater satellite, capable of receiving and retransmitting the equivalent of 600 voice-quality circuits for telephone, telegraph, data, facsimile, and television communications between ground terminal stations in the United States and in Europe. A few days later, on July 23, American and European television networks cooperated in telecasting the first intercontinental television program originated and seen by viewers simultaneously on both sides of the Atlantic Ocean.

A communications satellite system will provide not only greatly increased capacity to meet the future needs for international communications; it will also open up an unlimited opportunity for linking the world in a common dialogue, with far-ranging effects on the thinking, understanding and culture of all mankind. Even today, the beginnings of a change are apparent. When Sir Winston Churchill was made an honorary citizen of the United States, the great British statesman watched the ceremonies taking place in the White House rose garden in Washington on the screen of his own television receiver as he sat at home in London. At the same moment, many millions of viewers in the United States and in Europe, including a large number living behind the Iron Curtain in Hungary and Czechoslovakia, watched President Kennedy proclaim Sir Winston an honorary citizen.

This historic achievement was made possible by the communications satellite known as Relay I, a complex spacecraft launched by the National Aeronautics and Space Administration on December 13, 1962. Relay I had been designed and built for NASA by Radio Corporation of America, Inc., and it had capabilities broadly similar to Telstar I. However, because of its greater power, Relay I was able also to establish direct communications between ground terminal stations in the United States and in South America, which was not possible with Telstar I.

Telstar I and Relay I were of an entirely different configuration and internal structure, but they had two design elements in common. Both were low-level active satellites, which orbited from about a perigee of 600 miles to an apogee of 3,500 miles above the earth. Both were experimental research efforts and not prototypes of a commercial operational system. These satellites were launched to demonstrate that wideband radio signals can be relayed by active satellites and to measure the effects of Van Allen Belt radiation on various types of solar cells and other solid-state components operating in the space environment. Telstar I and Relay I obtained their power from silicon solar cells exposed over the surface of the spacecraft. Radiation-test devices measured deterioration of the cells and also other components while passing through the Van Allen Belt.

The miniaturization of electronic equipment had made it possible to include a large number of components and sub-assemblies within the satellites. Telstar I contained 15,000 components, including 3,600 solar cells and 2,528 diodes and transistors. It was a sphere of only 34½ inches in diameter and weighed about 170 pounds. Relay I had over 8,000 silicon cells alone and was only 27 inches at its base and 51 inches high over-all, weighing about 170 pounds. Yet both of these satellites contained intricate electronic equipment, including antennas, wideband transmitting and receiving equipment, traveling wave tubes, power supplies, and subsystems for tracking, telemetry, and command functions.

These satellites have been used to conduct hundreds of tests and have reported an extensive amount of scientific information concerning the space environment which was needed to develop an operational communications satellite system. Telstar I was operated successfully between July 10, 1962, and February 21, 1963. Telstar II was launched on May 7, 1963, and operated successfully for several weeks. Relay I, which was launched on December 13, 1962, operated successfully for a longer period than its predecessors. Relay II was launched January 21, 1964.

The ground terminal stations are of prime importance; several have already been constructed in the United States and in certain other countries. They are necessary to track the communications satellites and to transmit and receive the radio signals from them. The signal from the Telstar satellite at the ground receiver was only a trillionth of a watt. Thus it had to be received and amplified in the receiving system to be useful.

With the progress made thus far, it is clear that no scientific obstacle stands in the way of establishing a commercial communications satellite system. We are now on the threshold of using microwave techniques to establish a large-capacity global system. Microwaves are extremely short electromagnetic waves which have been used extensively since 1947 within the United States for large-volume transmission. Microwave transmissions travel only in straight lines, however, and because of the curvature of the earth, microwave relay stations are required some 20 to 30 miles apart in overland systems. Relay stations are not feasible over the long transoceanic distances, and so a system of communications satellites will serve as automatic microwave relay stations in space to receive and retransmit communications between widely separated points.

The operational altitude of a satellite and its inclination to the equator determine the duration of mutual visibility between any given pair of ground stations. If continuous world-wide communications were desired using satellites orbiting at an altitude of 1,000 miles, as many as 400 randomly distributed spacecraft would be required. By increasing the operational altitude to 6,000 miles, the number of satellites can be reduced to approximately 40. By further increasing the altitude to 22,300 miles, the number of satellites can be reduced to only three for global coverage.

Assuming a useful life in space of approximately two years, even 40 satellites represent an enormous investment, especially if the launching vehicles and facilities and the expensive and complex ground terminal stations are included in the cost. The use of three satellites at an altitude of 22,300 miles would significantly reduce that cost. They would rotate at an orbital speed exactly equal to the speed at which the earth rotates and would remain in a relatively fixed or synchronous position with the earth below. Equally spaced in equatorial orbit, they would provide coverage over the inhabited parts of the world with substantial overlap. At this altitude, each satellite would continuously see approximately one-third of the earth's surface. A satellite of this type placed at midpoint over the Atlantic Ocean would be visible to both North America and Europe, and continuous communications could take place between the two continents through a single satellite. Similarly, another satellite over the Pacific Ocean would provide constant communication between North America and Japan, Australia, and the Far East. A third synchronous satellite over the Indian Ocean would round out the coverage for messages between the Far East and the European and African nations.

Further testing and research will be necessary in placing synchronous satellites of large capacity 22,300 miles above the earth and maintaining them in proper position. On July 26, 1963, a research effort to explore the unresolved problems was made by NASA with the launching of Syncom II, an experimental satellite built by the Hughes Aircraft Company. The satellite weighed only 79 pounds and was designed to accommodate only one two-way telephone call at a time between North and South

America or between Europe and Africa or across the Atlantic. The successful launching of Syncom II was a major step forward. It proved the feasibility of a synchronous satellite orbiting with the earth's rotational pattern.

The Communications Satellite Corporation plans to launch an experimental satellite in synchronous orbit over the Atlantic Ocean in mid-1965. This satellite will be designed to provide up to 240 two-way telephone grade circuits between terminals in North America and Europe. Such circuits could be used for telephone or record traffic and for the transmission of television signals.

A synchronous satellite system offers great promise for the future. A system of this type, with multiple access and 1,000 or more duplex voice circuits, could be used simultaneously by many countries and operating agencies, employing their own ground stations to communicate directly with each other. The system would provide truly global coverage; it would be more economical, more efficient, and more flexible than any other known system. It would also require less complex ground stations, which is particularly desirable in seeking to extend satellite communications to underdeveloped nations.

Because the radio signals must travel long distances from one ground station to a synchronous satellite and onward to another ground station, there is a time lag of three-tenths of a second between transmission and reception. In a telephone conversation, this time interval will be doubled before an answering voice becomes audible at the first ground station. Experimental programs are under way to obtain information about public attitudes toward the time delay. This natural factor is of no consequence in the handling of one-way communications traffic.

A synchronous satellite system wtih multiple access is still in the future. But because our national interest requires the establishment, as expeditiously as possible, of a commercial communications satellite system, it is planned to begin operations with experimental operational satellites of lesser capabilities. Even so, an operational system of this type will make possible the exchange of television programs between major centers of population in the world. Eventually, in a global system, an event of importance in South Africa may be seen as it is happening by viewers in Japan, while viewers in Australia may be able to see an event in New York, Cairo, or Moscow. Television relays, however, will be only a fractional part of the traffic moved by the satellites. It is possible to foresee the world-wide storage of information, its retrieval and transmission, directed from centralized computer facilities and widely distributed memory systems. Status reports on major national or international problems may be transmitted from continent to continent. International communications facilities, as supplemented by satellite systems, can provide instant access to the pertinent laws, regulations, and procedures of any country from any other country.

The existence of a communications satellite system will also provide greatly expanded facilities for direct exchange by facsimile transmission of large volumes of business and personal correspondence between people widely separated by oceans and continents. Entire newspaper pages may be transmitted across the oceans for photographic reproduction and local printing. Alternatively, news copy originating in one continent may be word-counted and automatically justified for column-width by electronic computers and then set in type on another continent by keyboardless linecasters taking their computerized instructions via satellite. This technique has already been demonstrated successfully with existing low-level satellite communications.

Communications satellites can be used effectively with meteorological satellites for weather service. Since April 1960, RCA has designed and constructed a series of TIROS meteorological satellites, which have supplied essentially continuous television observation of weather phenomena on a global scale. These have clearly demonstrated the value of information-gathering weather satellites as a means of extending meteorological knowledge and of providing world-wide weather forecasting services.

The effectiveness of a global meteorological satellite system will depend upon the rapid dissemination of information and forecasts around the world. Specialized ground stations, including electronic computing and data-processing equipment, are required to guide weather experts in producing useful forecast information from the large volume of pictures and data transmitted from the meteorological satellites in successive orbits. A distribution system is required for the transmission of weather analysis and charts from the processing centers to other areas of the world while the information is still useful. Weather forecasts obtained by television observation can be passed quickly to any region around the globe by means of a communications satellite system.

Eventually, second-generation synchronous satellites may be operated with the use of atomic power supplies. In this manner adequate power may be obtained to operate several thousand two-way voice channels in each of the satellites. The higher power also will permit transmissions directly from the satellites with sufficient signal strength to be received directly in the home. This offers a powerful new tool for education, particularly in large underdeveloped areas where broadcast services are now lacking and are unlikely to be established in the foreseeable future.

It is well to bear in mind, however, that the orbiting of a few experimental communications satellites will not alone make a successful system. Even with the hundreds of successful experiments which have been carried out with the Telstar, Relay, and Syncom satellites, many more technical questions remain unanswered. What will be a reasonable life expectancy of communications satellites operating in the hostile space environment? What will be the effects of prolonged exposure to space radiation? Will damage to solar cells and other electronic components be appreciably less at high altitude, as compared with low levels where satellites must penetrate the Van Allen radiation belts? All of these questions, and many more, await further research and development with experimental spacecraft.

Complex and far-reaching international, economic, regulatory, legal, business, and administrative problems must also be resolved. In the United States, Congress has decreed that the development and establishment of a commercial communications satellite system will be within the framework of the Communications Satellite Act of 1962. The Satellite Act is a basic declaration of United States policy and purpose to establish, in conjunction and cooperation with other countries, as expeditiously as practicable, a commercial communications satellite system as part of an improved global network, which will be responsive to public needs and national objectives, which will serve the communications needs of the United States and other countries, and which will contribute to world peace and understanding.

The Congress declared that this policy was to be implemented through United States participation in a global satellite system in the form of a private corporation, subject to appropriate governmental coordination and regulation. The President is responsible for planning and development of the program, coordination of the activities of government agencies, and foreign participation and utilization of the satellite system for governmental purposes. The Department of State is authorized to render assistance as may be appropriate in foreign business negotiations with respect to facilities, operations, and services. NASA will advise on technical characteristics of the satellite system, cooperate and assist in research and development, and furnish satellite launching and other services on a reimbursable basis. The Federal Communications Commission is responsible for the regulation in the public interest of the establishment, operation, and use of the satellite system and of ground terminal stations. The Satellite Act also provides for the creation of the Communications Satellite Corporation, its organization, financing, powers, and purposes.

The Communications Satellite Corporation, incorporated on February 1, 1963, is carrying forward the program to establish a commercial satellite system. In mid-1964 the Corporation made its initial stock offering of $200,000,000. Fifty per cent of the stock was offered to the general public and fifty per cent to the communications carriers which were authorized by the Federal

Communications Commission to purchase such stock. This stock offering was oversubscribed by the public and the carriers, indicating confidence in and support for the future of space communications.

The Communications Satellite Corporation is authorized to plan, initiate, construct, own, manage, and operate the system in conjunction with foreign governments and business entities. And the Satellite Act provides that the President shall "insure that timely arrangements are made under which there can be foreign participation in the establishment and use of the system." International aspects of the program, however, involve the intricacies of foreign relations and are among the most difficult problems to be resolved. In order to establish the system in conjunction with other countries, negotiations and agreements must be reached on many matters, including frequency allocations, the technical specifications, responsibilities, capital contributions, the nature of the system and the extent of its use. The United States international carriers will also be concerned with foreign negotiations in regard to their use of satellites facilities, operations, and services.

Frequency allocations for communications satellite service were made at the Extraordinary Administrative Radio Conference on Space Communications held in Geneva, Switzerland, in October and November 1963. This conference was held under the auspices of the International Telecommunication Union, a specialized agency of the United Nations.

In July, 1964, eighteen countries concluded interim arrangements covering commercial, technical, and financial aspects for the establishment and operation of a global communications satellite system. These countries were the United States, Australia, Austria, Belgium, Canada, Denmark, France, West Germany, Ireland, Italy, Japan, the Netherlands, Norway, Portugal, Spain, Sweden, Switzerland, and Vatican City. These arrangements will be open for six months on the same terms to the other countries which are members of the International Telecommunications Union. Only Communist China and North Korea are not members.

In the United States there are a number of competing international communications carriers. Several of them have indicated a desire to own and operate their own ground stations for satellite communications. This is a matter of international carriers providing their own transmission facilities wherever feasible to serve their customers better and also participating to a greater extent in the revenues from international communications.

The Satellite Act authorized the Federal Communications Commission to grant appropriate authorizations for the construction and operation of ground stations, either to the Communications Satellite Corporation or to one or more authorized carriers or to the Corporation and one or more such carriers jointly, "as will best serve the public interest, convenience and necessity." Thus, the determination of the pattern of ground station construction and operation in the United States is an important question for the future. In foreign countries, the ground terminal stations will be established by the governmental administrations and business entities providing international communications services from such countries. Although ground stations have already been provided for by England, France, Germany, Italy, Brazil, and Japan, the construction of a world-wide network is still far from complete.

Substantial economic problems arise in establishing an operational commercial satellite system. The capital requirements over a ten-year period have been estimated from about $350 million to $650 million for a synchronous satellite system and $550 million to $1.4 billion for a medium-altitude system. The direct carrying charges have been estimated at $40 million to $80 million annually for a synchronous system and about double these amounts for a medium-altitude system. General administrative expenses and return on investment are added revenue requirements.

However, the international communications industry, which will use the commercial satellite system, is relatively small. In 1963, the international communications revenues of all United

States carries amounted to only approximately $175 million. This compares in the same year with over $9 billion of United States domestic communications revenues. Moreover, it is anticipated that present and planned facilities will be adequate to handle the volume of international traffic expected through 1965.

A commercial satellite system will help handle the future growth in international communications traffic. But a communications satellite system will not be the exclusive method in the future of international communications. Today, international communications services are provided by submarine cables and high-frequency radio systems. These facilities already constitute an efficient and world-wide system. Additional cables are being constructed between major traffic centers of the world each with capacity up to 128 simultaneous voice-quality circuits. A transistorized submarine cable, which is being developed by American Telephone and Telegraph Company, will have transmission capacity up to 720 voice-quality circuits and can handle international television. The transistorized cable may be available by 1966.

The uncertainty of demand for satellite communications services, coupled with the high capital and operating costs of a global satellite system, raise the question of how soon the Communications Satellite Corporation, a private enterprise, can reasonably be expected to operate profitably. The late President Kennedy indicated early in 1962 that there "may be quite a long period of time before there is any return on this investment." The Communications Satellite Corporation has indicated that it may operate at a loss for several years even after commencement of full operations.

Thus many challenging problems must be resolved to bring into being a successful global communications satellite system. Its ultimate success in both performance and profitability as a private enterprise will depend upon numerous factors, including the nature and extent of participation and cooperation by foreign communications agencies, the nature and costs of further progress in space technology, the kind of satellite system which will be established, and the growth in the demand for international communications services.

Many of the problems of satellite communications are unique and without precedent. Through the coordinated and concentrated efforts of both government and industry and the cooperation of other nations, they can be nonetheless resolved so that this foremost peaceful application of space technology will serve to benefit all mankind.

SPACE AGE COMMUNICATIONS: A BETTER VIEW TO UNDERSTANDING

James C. Hagerty

The art of communications, which has grown so rapidly in the first half of this century, has been enormously accelerated since the late 1950's by new technologies from space research. All forms of communications have benefited; but television, the youngest of our mass communications media, has probably shown the greatest gains. These will increase and television will acquire even broader dimension when an operational world-wide communications satellite system is developed.

Global communications by satellite systems will give television greater influence than that of other mass communications media because of television's uniquely personal nature. It brings persons and events directly into our homes. Its pictorial view, with sound added to sight, has a much more profound impact than that conveyed by the inaudible words on a printed page or just the sound of radio. Television images cut across language barriers and, to a certain extent, political barriers, for they strikingly show one people what another is doing. Television is like a two-way street where we are able to see others and, in turn, be seen by them.

In the eighteenth century, the Scottish poet Robert Burns wrote:

> Oh wad some power the giftie gie us
> To see oursels as others see us!
> It wad frae monie a blunder free us,
> An' foolish notion.

The gift of self-sight, "To see oursels as others see us!" is now nearer and within reach for our nation and others through the clear eye of television. This self-sight may not always save us from blunders, but it may force their recognition. Facing a fault is often a first step toward its correction. When, for example, the television camera showed a restaurant owner in Cambridge, Maryland, just how he looked when he abused Negroes who attempted to enter his segregated establishment, he confessed that it made him feel ashamed.

The prospect of instantaneous world-wide television and our traditional support of a free press makes it imperative that individually and collectively we put our best face as well as our best foot forward. Whether or not we would want to hide our faults, if we could, the fact is that our open society denies us this strategic comfort. It is well to recall that the one person killed in the Oxford, Mississippi race riots when James Meredith was enrolling in "Ole Miss" was a foreign reporter. We can be sure that when satellite telecommunications are internationally available, foreign newsman will be here to record the picture of America, whatever it may be, for the world to see.

Our leadership in world affairs in no small measure depends upon the image of our democracy. We, as citizens of the United States, therefore, have a grave responsibility to determine the image we present to the world—the deed as well as the purpose and intent to which we give voice. One has only to travel road in countries where the majority of citizens have colored skins to appreciate how essential this is. On the advance trips I made as

Press Secretary to President Eisenhower to countries in Africa, Asia and the Far East, the most difficult job I had in social, after-business-hours conversations with my opposite numbers was to try to explain how Americans were attempting to solve the racial problems of the United States which often were greatly exaggerated in their newspapers. The cold printed word or even the still photograph, effective though they may be, do not have the tremendous emotional shock of televised sight and sound.

If the electronic eye accents or magnifies the weaknesses of our democracy, it also emphasizes its unique strength, as no other mass communications media can. Could printed words, however eloquent, convey the majestic dignity of the thousands of our citizens who gathered in the nation's capital in the late summer of 1963 to demonstrate for jobs and freedom? The films that were telecast by nationwide networks within the United States and transmitted to Europe by a Telstar communications satellite portrayed much more than a mass protest against inequality. Audiences everywhere could also see that the right to petition for human rights was recognized and supported at the highest levels of our national government. I am certain this fact was not lost on Communist viewers if they were permitted to see them.

The Russians were first in space with machines, animals, and men; but no foreign reporters, either from newspapers or from radio and television, were allowed behind their Iron Curtain to report the Soviet space exploits. Every NASA launch, on the other hand, has been open to the press of the world. Our failures as well as our achievements in space have thus been exposed to public view. If the televised failures reflected some technical weaknesses in the space program, they also were evidence of a nation morally and intellectually strong enough to show error and confident enough of ultimate success to avoid trying to justify the failure. I am certain that the complete coverage of all of our manned space flights from launch to recovery significantly lessened the effect abroad of the spectacular but secret Soviet manned space experiments. But static self-satisfaction over past accomplishments is not a luxury enjoyed by a free society, for it can neither hide nor ignore failures and blemishes. This constant exposure, like bitter but beneficial medicine, may often be hard to take, but it is good for us. For the only way we as a democratic people can avoid public focus on our wrongs is to right them.

Since June, 1962, the United States has launched a total of six active experimental communications satellites: two Telstars, two Relays and two Syncoms. With the exception of Syncom I, all have operated successfully, with clear transmission of telephone, teletype, wireless, radio and television. They have enabled audiences in Europe and the United States to view simultaneously some of the great events of our time, such as a Presidential press conference in the auditorium of our State Department; the late President Kennedy's 1963 visit to Europe; the death of Pope John XXIII and the election of Pope Paul VI. Our satellites have also transmitted the first telecast by the National Broadcasting Company of an international art program, a museum without walls, in which art treasures were exhibited directly from the Louvre in Paris and from the National Gallery in Washington. Ron Cochran's regular daily news program on the American Broadcasting Company's television network originated in Europe and was sent to the United States by Telstar for several weeks during the late President Kennedy's last trip to the European continent.

Through satellite telecommunications we have been able to exchange viewpoints as well as views. A Columbia Broadcasting System (CBS) "Town Meeting of the World" via satellite featured an interchange of opinions and observations on international questions between former President Dwight D. Eisenhower speaking from Denver; former British Prime Minister Sir Anthony Eden from London; Henrich von Brentano, majority leader of the West German Bundestag, from Bonn; and Jean Monnet, France's "Father of the Common Market," from Brussels. Another CBS "Town Meeting of the World" had a discussion on the Christian Revolution with church leaders in London,

New York and Rome participating. An international medical consultation was achieved via satellite when an electroencephalogram, as it was being taken of a patient, was transmitted from a hospital in Bristol, England to neurologists attending a meeting in Minneapolis, Minnesota of the National Academy of Neurology. A diagnosis was made by the American specialists and reported via satellite back to Bristol.

Telstar I, our first communications satellite, relayed the sound and pictures of a meeting of the East German Communist Party Congress to audiences in the United States in January, 1963. American television, via Telstar II, was brought live into the Soviet Union for the first time on November 26, 1963, with reports of the tragic event of President Kennedy's assassination and his burial in Arlington National Cemetery before the largest gathering of foreign dignitaries in history. Soviet editorial comment in Izvestia noted that Telstar, "the technical wonder of the twentieth century, came into our lives with America's mourning." It added, "We have seen the grief of the American nation and profoundly sympathize with it."

The value and importance of an operational communications satellite system was recognized by the Congress which enacted legislation, signed into law by President Kennedy, for the establishment of a Communications Satellite Corporation (COMSAT). The law provides that COMSAT be a private corporation and stipulates that its stock may be purchased both by the general public and private corporations, and the funds used to develop a communications satellite system. COMSAT has already applied to the Federal Communications Commission for authority to launch a synchronous-orbit satellite over the Atlantic Ocean early in 1965 for experimental purposes, with a launching of a satellite for operational use to follow in 1966. The first stock offer made in June, 1964, was well received even though COMSAT's prospectus warned that the venture is likely to be high in risks and low in profits and dividends for some time to come.

Yet mingled with the strong support for COMSAT, there have been some reservations and doubts expressed, not so much over securities as over security. There have been suggestions that a satellite communications system will overexpose us to foreign propaganda and that this might be counteracted only with a system owned and operated by the government. These views stem largely from a misunderstanding of the function of government in communications and from a misconception of how a communications satellite will operate.

Radio and television stations are licensed by the Federal Communications Commission. They are regulated, not controlled or owned by the government. This is both proper and necessary, since the technical nature of radio and television requires assignment or licensing of channels to prevent chaos in transmission and reception. Program content is left to the judgment of the networks and the stations and, ultimately, public opinion. Under our democratic system, government cannot be allowed the power to interfere with the movement of news and news copy; nor is it to be given the authority to assume the position of cultural arbiter or censor.

The threat of a sudden influx of foreign propaganda into American homes via satellite communications has no basis in reality. It is based on the incorrect assumption that programs will be beamed directly from the satellite into our living rooms. This is not so. All program material would first come into the networks for dissemination nationally to local stations. The networks will in no way surrender their editorial judgment, and I am sure we will continue to be wise in the exercise of that judgment. Our television staffs are well informed and highly trained in their professions; and as long as they continue to be informed, there is little danger of misinformation or propaganda—either foreign, or domestic. I would say again that it is what others may see of us rather than what they may show us that should engage our concern.

In this connection, it has been said semifacetiously that the worst propaganda that could be used against us is of our own making and is commercial rather than political. Television and radio advertising, according to some critics, convey an image of

Americans as a people who suffer from excessive body odor, bad breath, dandruff, stuffed-up sinuses, chronic headaches, stomach disorders, whose women need padded brassieres and whose men depend for their appeal on hair lotions. I disagree. Those much maligned commercials keep television free and contribute greatly to our national economic life.

But I must admit that when Soviet newsmen and television producers have visited here, they have been openly amused by our commercials, particularly those concerned with brassieres and remedies for stomach disorders. But commercials are a part of the American way; and, very frankly, after watching Soviet television during my visits to Russia, I am not so certain that our commercials might not be a welcome diversion to the Russians and even appear more attractive than the constant barrage of political commercials to which they are exposed. The citizen behind the Iron Curtain might find the Madison Avenue line much more to his liking than the Red Party line.

However, I doubt very much that our special brand of commercials is an exportable product. They are designed to sell American products and would hardly be useful in countries where such products are not widely sold. In any event, neither our commercials nor our regular entertainment programs would be moved over communications satellites. They can be transmitted far more reasonably by plane. For some time to come, the high cost of satellites will restrict their use to the transmission of special events of broad interest and significance and history in the making, as far as television is concerned.

Space research has also brought other developments to telecommunications. Though not as dramatic as the communications satellite, they are nevertheless important to the advancement of these media. For example, miniaturization techniques from our space program have made it possible to reduce television equipment in size, weight, and cost. Taping machines weighing less than 100 pounds are replacing the costly two-ton giants used for telecasts. At the American Broadcasting Company, our engineers have been using miniature taping machines which weigh only 60 pounds and which are only a little larger than a valise. They cost about one-third of the $45,000 that each of their big brothers costs. These miniature taping machines were used successfuly at the 1964 Winter Olympics and the 1964 political conventions.

Television cameras are being scaled down so that they can be carried with ease. In cooperation with the Sylvania Company, ABC has developed a miniature five-pound television camera. This camera needs only a twenty-three-pound pack which can easily be strapped to the back of the camera operator and functions perfectly. Without attached wires or cables, the image-signals from this five-pound camera can be transmitted up to a mile's distance and then piped into portable truck equipment or stations. In other words, television equipment is beginning to give the television reporter the same mobility in his coverage of events as the newspaper reporter who has only his pad and pencil to encumber him.

The expanding influence of television does not mean that it can or will supplant the other mass communications media. Newspapers, magazines, and radio are each an important part of communications. Each is complementary rather than competitive in its relation to the others. Each has a special advantage. Television has the advantage of showing what is happening at the time it occurs. But it is limited in how much it can show by the second hand moving on the clock. Radio is similarly limited. The printed word of the daily newspaper can overcome the restrictions of time because it can increase space. The daily newspaper can go into far greater detail than either television or radio on the events of the day that happen here and abroad. The periodical can take both time and space to analyze what television may show only briefly and what the papers may report in greater length but in lesser depth.

The media of the printed word also stand to gain from global communications satellites and new Space Age technologies. Dispatches will be accomplished at fantastic speeds. For example, in less than time than it would take to dictate a sentence, a full page of copy can be flashed overseas and reproduced electronically on paper. It is not impossible that this and other new developments will speed the trend that is just beginning in the United States—the establishment of national newspapers. None of our papers is really circulated nation-wide as is the *London Times* or the *Manchester Guardian* in the United Kingdom. Television and radio networks now meet this need by handling national and international news and calling upon affiliated stations for local and regional coverage.

The next step from national daily publications could be international or "orbital newspapers," which global communications satellite systems may make possible. Language would pose no barrier, for computers could be tied into the systems, providing almost instantaneous translations for each country. Electronic presses would print the international editions for distribution, all within a matter of a few hours.

It appears certain that the scientific and technological advances, present and future, in the field of communications will make us the most informed people in history; but will we be the best informed? We can if we will only take the time and effort to avail ourselves of the abundance of information that will be readily ours and, with thought and care, put it to good use. For communication means more than transmitting and receiving information. Among the definitions given for the word *communicate* is "making common to all that which is given usually by one or to one . . . sharing with another or others what is primarily one's own." If we as individuals and nations can learn to communicate in this sense, we will have achieved that community of understanding that has thus far eluded mankind.

SPACE AND LANGUAGE

Richard B. France

In the development of technology, language as a means of communication serves as a catalyst; yet unlike most catalysts it has not remained unchanged in the process. The technological advances which gave birth to the Space Age have not only doubled the body of our scientific knowledge in twenty years but have added dimension and color to our language. Barely have we become accustomed to the language of the automotive and jet age when we are exposed to such terminology of the Space Age as astrionics, escape velocity, apogee and perigee, specific impulse, zero-g, lox, and many others.

The National Aeronautics and Space Administration's *Dictionary of Space Terms* now has some 80,000 entries; and glossaries by the Air Force and industry are adding to these. Many of these words and phrases already have become a part of our everyday language. Such phrases as "go-no-go," "A-OK," and "all systems green" are enriching our idiomatic speech; but other words and phrases coined by those in the space field are part of a private language which can communicate only confusion to the eavesdropping layman.

PERT ANNA, for example, in space lingo does not refer to a saucy girl. PERT is the acronym for Program Evaluation Reporting Technique, a management system applied to space projects. ANNA is the acronym for an Army-Navy-NASA-Air Force geodetic satellite. Used together the words mean the management of a particular satellite which has resulted from the coordinated planning of many departments of government.

You could reasonably expect "angels" in space; and they do, in fact, exist. It is, however, somewhat of a shock to discover that HADES EMPIRE has a place in the heavens and that SATAN, as overseer, is up there too—as well as elsewhere. "Angels" in space are radar echoes caused by a physical phenomenon not discernible to the eye. HADES is a Hypersonic Air Data Entry System that could be useful for EMPIRE, an Early Manner Planetary-Interplanetary Round-Trip Experiment, which SATAN, a Space Automatic Tracking Antenna, may keep under surveillance.

In the language of space, POGO does not refer to the familiar comic-strip character; it is a Polar Orbiting Geophysical Observatory, part of a family of geophysical satellites that include OGO, a standardized Orbiting Geophysical Observatory, and EGO, an Eccentric-Orbiting Geophysical Observatory. Romping in the far reaches of outer space with NERV (Nuclear Emulsion Recovery Vehicle) and SNAP (System for Nuclear Auxiliary Power) are IMPS (Interplanetary Monitoring Probes), NUDETS (Nuclear Detection Systems) and SPADATS (Space Detection and Tracking Systems). In the programs and hardware which these acronyms represent, the nation has, in any language, a capital asset in space. In space talk, however, ASSET is an Aerothermodynamic-elastic Structural System for Environmental Testing, used to check out re-entry effects on winged space vehicles.

Even ordinary words get new meaning from space. The word "rocket," which the layman may think he understands, is not a term of uniform meaning in space use. A rocket becomes a missile when it is designed to carry warheads. Those which place men or scientific experiments into orbit are launch vehicles. To abort in space means to cancel or cut short a flight.

Space also has been the inspiration for the creation of new words. One of these is *terrella,* Latin for "little earth." This is a term coined by Dr. Hubertus Strughold, known as the "Father of Space Medicine," for the closed environmental system in the space cabin by which man is sustained in flight. The term is apt, for man must take his earth environment with him if he is to survive in space.

Dr. Strughold also may come to be known as the father of space language for the many words he has coined and which are now in international dictionaries. These include *Bioastronautics* or the art and science of the construction of spacecraft and life support systems which have enabled man to travel in the *Gravisphere,* or area of predominate gravitational attraction within which a satellite can be held in orbit. If we are to find life in outer space, Dr. Strughold says it will be within the *haleoeccosphere* or zone around a sun in which the radiant conditions are favorable for life on planets moving within this zone. He has named the coming era in geological history, which we have called the Space Age, the *dosmozoicum* (Age of Life in Space).

Some English purists, highly critical of this new terminology, believe that the language of space, if *not* out of this world—ought to be. To them it is a nonlanguage, where neither words nor their meanings are what they seem or should be. They tend to overlook the fact that the language of space—like all languages—develops as it is needed. The lexicon of space terms has grown from the necessity of naming new materials and concepts, and out of the very real need for scientists and technicians to achieve a common basis of understanding and a means for communicating special knowledge. It is a kind of science interlingua which cuts across all science disciplines.

Space experiments require a variety of specialists. For Project Mercury, specialists in nearly a hundred fields were needed.

Effective interdisciplinary communication depended largely upon common terminology. New words and phrases were coined, often by combining terms understood by the several disciplines, to provide the required language link fundamental to communication. The language of space, which may be a babel to the laymen, is the self-created means of communication of specialists in science and technology.

Technology today is moving ahead every four years as much as in the hundred years prior to 1945. Thirty million technical books are already on library shelves and six hundred more are added daily. About 100,000 special journals are published yearly and 35 per cent of them—with over three million articles—deal exclusively with science and engineering. Even the slow-paced turtle is the subject of nearly three hundred books and articles each year. It is estimated that doctors would need to read an average of one book per hour just to keep abreast of the literature in their field. It is very easy for information to get lost or obscured in this huge stockpiling of human knowledge. Its transmission is a challenge recognized by both government and private organizations.

Unable to meet this challenge by ordinary means of communication, new methods for sorting, filing and retrieving known data have been devised as well as ways to accelerate their transmission. To keep pace, language—as a tool of communication—has added new forms other than the spoken or written word. These are encompassed in the computer, which has proved to be an invaluable tool in storing and retrieving data.

There are several ways of deriving and assigning terms for indexing material; but to find a document in an information retrieval system requires the same procedures which were used originally to catalog and store the documents.

There are three broad systems for storing, and hence retrieving, documents. In one, the material is filed by subject matter, as in most offices. In the second, the search focuses on descriptions of the items in a collection rather than on the documents themselves. An example of this is the familiar library card catalog index which describes a document and gives its physical location. In a third system, a microfilm of the document is filed physically with a coded description of the material. In all these systems, computers have been used to speed the matching of the information sought with descriptions of documents in a collection. Coupled with other devices, computers have provided the researcher with a list of relevant documents or abstracts of the material contained in pertinent documents.

The use of the computer as an implement to research is growing rapidly. At least one large library system, that in New York, is considering converting to modern methods of information retrieval and may very well set the pattern for other libraries. It has been suggested that several large library systems strategically located could rapidly make available an infinitely greater amount of information to its subscribers by processing requests and mailing either documents or microfilm images or abstracts.

The Pentagon's Defense Documentation Center plans to produce microfilm of all the documents in its collection and keep the film on file in New York City, in Dayton, Ohio, and in San Francisco and Los Angeles. This will enable a defense contractor to have the search for data performed at the Defense Documentation Center in Arlington, Virginia, then view the document films at the center nearest him and obtain the prints, if necessary. From its current store of nearly 650,000 documents, the Defense Documentation Center processes over three-quarters of a million requests for information per year.

Computers can do more than deliver messages or find source material. They also can answer questions. Space vehicles and rockets have been "flown" by computers before they were built; and after they have been built and launched into orbit, they have been monitored by the machines. The Air Force has pioneered in using electronic data-processing equipment to control inventory levels and speed its flow of supplies to bases around the world. Many of its methods have been successfully copied for industrial use. In industry, computers direct and guide machine tools in manufacturing processes, watch stock levels and prepare reorder forms, pay personnel, and digest, analyze, and transmit thousands of bits of data upon which management may base business decisions. Their language is a vital communications link for our era.

The computer speaks in code, using the alphabet or numbers or other symbols punched on cards or paper or inscribed on magnetic tape. Using these codes, the human programmer translates instructions or questions to data-processing machines and the machine responds with the "readout" or answers. At the present time, each computer system has its own language, and the interchange of information between different makes of machines is possible only after costly and time-consuming translation. However, the American Standards Association, with the support of members of the computer industry, has developed a new alphabetical and numerical language so that all makes of electronic computers and data-processing machines can talk with each other. The magnitude of this interlingua for machines is indicated by the fact that $3 million worth of man hours were expended over a four-year period to achieve this standardization. The new standard code will speed operations from tracking of missiles and satellites to routine ordering of supplies.

The expansion of computer use, which will be accelerated by this standardization of its language, has created the need for new circuits for machines to talk to machines in exchanging business and defense data. This aspect of communication has grown so rapidly that the American Telephone and Telegraph Company now estimates that machines will become our biggest conversationalists and that in another ten years revenue from this source of traffic will outstrip that charged for human conversation. Great strides have been made in these processes and techniques; and in the main, they have kept pace with the rapid growth of technology. But do serious gaps remain and what is the shape of the future?

It has been estimated that scientists now spend half their time searching through indexes and libraries to try to find data on earlier efforts in areas in which they are working. If this search time is to be reduced by information retrieval systems, we must develop greater precision in the definition of words. The gap between a question and documented information must be bridged by language of precise and universal meaning. Here we could stand aid rather than criticism from our linguistic specialists—our language purists. In essence, the scientist today must add to his vocabulary the language of the information retrieval system he uses if he is to find the answers to his queries.

If scientists of various disciplines have difficulty communicating with one another using a common language, the problems are multiplied when they are confronted with a foreign language. Foreign technical papers must be translated, and terms which have a single connotation in one language may have multiple meanings in another or defy translation completely. Some experts, however, do not view the language barrier as serious, since most of the free world is bilingual with English emerging as a common language at international scientific meetings. A common awareness of English will probably minimize the need for translations in the future. Meanwhile, computers have been applied to translations. A computer-translator, like the one which scans the numbers on bank checks for information, is reading foreign words and printing out rudimentary English equivalents. Computer translations are not yet precise and inclusive. but they do provide an abstract of the material which is sufficient for a grasp of the content and the basis for a decision on whether a fuller, more meaningful translation by a human linguist is warranted. The promise of almost instant computer translations plus the adoption of a common language, possibly English, as a universal scientific tongue has thus reduced the problem of multiple languages as a barrier to communication.

There has been another force at work to disseminate information about the activities of scientists. Since Sputnik there has been a shifting of news emphasis to science. Trained reporters of the mass media are interpreting the progress of science and the meaning of our national space programs for the public. And here again, language is the key; for the art of science reporting today is to translate the technical and complex aspects of scientific advances in nontechnical language without havoc to the science. Subjects which are generally abstruse are not made clearer by turgid language limited to the terminology of the specialists working in the field.

The importance of explaining science and engineering developments in clear and concise language is recognized today by scientists and engineers. At a meeting of the American Institute of Mining, Metallurgical and Petroleum Engineers early in 1964, the need for more effective communication was discussed informally by many of the 2,500 persons attending. There was general agreement with the view expressed by Madan M. Singh, assistant professor of mining at Penn State, that unless a technical idea could be expressed clearly enough to be sold to management, it was no good. Thus it would appear that understanding and financial support, necessary for science and engineering as for every other human endeavor, will insure that the language of science makes sense. Without simplification and interpretation, the average person would find it difficult to keep pace with the terminology of the "Dosmozoicum Age" now just beginning.

Scientific advances in the Space Age have increased man's knowledge of the universe and given him a better understanding of himself and his environment. It remains to be seen whether or not he can, through the power of words, use this knowledge in the rational solution of his problems.

THE MORAL DILEMMA OF THE SPACE AGE

Abraham J. Heschel *

We are in an era of great scientific and technological advancement. Science has pried from the atom forces of tremendous power that hold both promise and peril for mankind; technology has developed vehicles that can be used to destroy the earth or search out the secrets of the universe. In this Space Age, to a greater degree than ever before, we all face the dilemma expressed by Moses: "I have put before you life and death, blessing and curse. Choose life."

Man has been endowed by God with the greatest and most awesome of freedoms: the freedom of choice. The deeper meaning of this freedom, applied to our programs in space, compels us to consider their political and social consequences from a moral perspective. The choices presented pose their greatest challenge to religion—a moral challenge, which is not being met. This goes beyond the question most often put; namely, whether space exploration is against God's will. There is no commandment

* Based on an interview with Lillian Levy.

or prohibition forbidding the exploration of space. But the fact that an action is not forbidden does not necessarily justify its pursuit. Nor does it make it morally right. The deeper meaning of the gift of free choice is that it is not always a question of choosing between good and evil, but also of choosing between two good things.

It is from this perspective that I question the enthusiasm with which many people rejoice in our scientific and technological achievements in space. Some religious leaders have praised them as a great triumph of the spirit of man. A more realistic appraisal is that the triumph belongs to science and technology, which threatens the enslavement of the spirit of man by inhibiting freedom of choice. For the sheer dynamics of modern scientific and technological developments interfere with the human capacity for decision. This would be true even if our only objective in space was to discover truths about the universe, which it is not. We are exploring space, not so much to seek scientific truths or because we are motivated by ennobling philosophic insight, but largely because space exploration has political and military value for the state.

In the past, science was subservient to the church. Its emancipation from the church—its freedom from the dominance of religious dogma—was a cause of pride and celebration. Now, for the first time in history, science has become the handmaiden of the state. Now, science must satisfy the demand of the state, and that demand is power. Therein lies the danger of its secular subservience and the cause of its conflict with humanity. For power, even if prompted by moral objectives, tends to become self-justifying, and creates moral imperatives of its own.

Our objective in advancing our capability in space is, as stated by our political leaders, to demonstrate power that will assure the security and maintenance of our democratic institutions. On this basis, the conquest of space has been given top priority; and a great portion of the nation's wealth and talent is dedicated to such programs as manned lunar exploration and the search for extraterrestrial life.

Some of our most thoughtful scientists allege that the discovery of life elsewhere in the universe would be the most important and rewarding achievement of mankind. If I question this, it is not because such discovery would not be in harmony with our belief in God. On the contrary, if God in creating the universe has created life elsewhere than on our planet earth, this is in perfect harmony with the Judaic understanding of the might and wisdom of God. I challenge the high value placed on the search for extraterrestrial life only because it is being made at the expense of life and humanity here on earth. It is on this basis, too, that I object to the high price being paid for manned lunar exploration.

Is the discovery of some form of life on Mars or Venus or man's conquest of the moon really as important to humanity as the conquest of poverty, disease, prejudice, and superstition? Of what value will it be to land a few men on the wilderness of the moon if we neglect the needs of millions of men on earth? The conflict we face is between the exploration of space and the more basic needs of the human race. In their contributions to its resolution, religious leaders and teachers have an obligation to challenge the dominance of science over human affairs. They must defy the establishment of science as God. It is an instrument of God which we must not permit to be misused.

It is our duty to point out that in placing lunar exploration above more fundamental human values there is a loss of self-respect, a sort of cheapening of human life. While there is no theological prohibition against doing research beyond the confines of this planet, what is really involved is the matter of doing the right thing at the right time. In my judgment, this is not the right time to invest more than $5 billion annually in space, measured against the less than half a billion we are allotting over a four-year period for the retraining of men and women whose productive capacities have been made obsolete by a mushrooming technology, or the less than $1 billion that President Johnson has recommended to fight the poverty that keeps more than one-fourth of our nation ill-clothed, ill-housed, and ill-fed. On any moral or ethical basis, when we can overlook the suffering of humanity in our childish delight in our ability to place monkeys and men in orbit around the earth, we are ill-prepared spiritually and morally for the vast accumulation of power which we are achieving through science.

God has a stake in the life of every man. He never exposes humanity to a challenge without giving humanity the spiritual power to face that challenge. Admittedly the challenge today is enormously great. We live in a time when we are going through several revolutions simultaneously: political, social, scientific, technological, and spatial. This has never happened in history before. But we must exert the spiritual will to focus the attention of our minds and hearts on the problems we face. We cannot avoid them by reaching for the moon or grasping for life elsewhere. We must turn our efforts to rediscovering the true value and dignity of man, what man's life means as a totality in its great dimensions, his great potential for the creative arts, for the advancement of science in the search for peace and understanding, for acts of charity.

The tragedy of man is that he is so great and that he fails to recognize his greatness. Jean Paul Sartre has said, "Man is condemned to be free." God has given him "choice"—the greatest obligation of freedom. He is waiting for man to exercise that choice. Man no longer can afford to compromise by accommodations not premised on moral values. The choice for man and humanity in this Space Age lies not in the stars but right here on this blessed planet earth. Will mankind fulfill its great destiny? Who can predict? We can only hope, pray, and demand.

THE HIGHER PROMISE OF SPACE EXPLORATION

Francis J. Heyden

The Space Age has come almost providentially to men on earth; for it has prompted an intellectual war, a competitive effort among men and nations to win primacy in the exploration of the universe, that holds manifold promise for mankind. In this competition, science and technology have been applied to the development of rockets and satellites, not to destroy men but to overcome their common enemy: ignorance.

Each victory in space is yielding such an abundance of knowledge and scientific data that the cooperative efforts of all are needed for collection, analysis, and productive application. No country on the surface of the earth can observe its own satellites unaided by other countries. Satellites pass over the entire world, and it is only the world as a whole that can study them. The united effort this requires gives promise of bringing both better understanding between men and nations on earth as well as a greater understanding to man on earth of the marvels of the universe that surrounds him.

The greater promise of the Space Age is that it will bring mankind closer to true fulfillment through God. Man's position with respect to God is primarily one of an intelligent being who is able to know, understand and appreciate what God has given him. The more we know of the universe, the more we know of the works He has given and done. This brings us closer to God and closer to that perfect fulfillment: to know God. Space exploration is a means to this divine achievement. In the psalm of King David, it is written, "The Heavens tell of Thy Glory, O God." Through science in space, this Glory may be revealed more fully.

Among these revelations of God's manifold works may be the discovery of worlds other than the earth on which we live and the existence of intelligent beings who may have a more perfect life and live in greater harmony with nature and themselves. Scientific probabilities suggest the existence of other worlds, and it is, indeed, possible that our earth is not, as we would like to believe, "the best of all possible worlds."

Fear has been voiced that the existence of such beings, if demonstrated, would be in conflict with present concepts of theology and that, therefore, the search for extraterrestrial life is contrary to the will of God. Some have suggested that we may need a new theology for outer space. But I can see no conflict between science and religion, between space exploration and the will of God. God's house is very big and man can explore it to the utmost. Not one of the Ten Commandments given to us says, "Thou shalt not travel to other worlds," or orders us to cover our eyes before the wonders of nature. We say that God is everywhere. Certainly, that is comprehensive enough to cover the most remote regions of outer space.

The concept of more perfect beings forms no difficulty in my mind. Indeed, I hope more perfect beings than we may exist. If such beings were to arrive on our planet, I would welcome them and hope that their intellect, brighter and less clouded than ours, would help us solve the problems of the human struggle. Such beings would be a reaffirmation rather than a denial of God, whether they were shaped like potatoes with the eyes of the potato or were two-headed and multi-limbed; for if they have intelligence, they know and appreciate the existence of God. They may not have had a revelation such as we have had here on earth; but whether or not God would have revealed Himself in the same special way He did for us, His existence would be known; and they would follow the same moral order.

St. Paul has said that there is a law in the heart of man which runs parallel with the commandments that have been given to us by revelation. Even the uncivilized savage knows certain moral fundamentals: that murder is wrong; that it is wrong to steal; that there is a Supreme Being that punishes evil and rewards good. This moral order, recognized by primitive tribes in the remote corners of our earth, would, I am certain, be known to the hearts of more intelligent non-earth beings.

Both as a scientist and a clergyman, I find nothing to fear from the scientific discoveries that have led to our exploration of outer space or that may result from it. History has shown that such fears are themselves a danger and, if indulged, may hinder both the spiritual as well as the material advancement of mankind.

It was not so long ago that attempts by scientists to find cures for certain diseases were opposed violently because of the mistaken conviction that God did not want them to be cured. When cures were discovered, opposition on theological grounds vanished. Man generally reconciles his concept of theology with what he comes to know and understand; for it is not possible for religion to contradict the testimony that God reveals in the wonders of the universe.

Great scientific discoveries may arouse opposition and inspire fear because of their possible destructive use. The terrible military potential of atomic power has led some to say that God should have put a barrier to atomic research. Yet, the atom holds as much promise for mankind from its use for peaceful purpose as it does danger from its destructive application. There are those who see in the satellite the specter of an orbiting arsenal from which bombs could be dropped anywhere on earth. The slogan that "He who rules the moon rules the world" is already familiar. It still is beyond the ability of our best scientists to develop the necessary accuracy to pinpoint a target on earth from a satellite traveling in orbit at a speed of more than 18,000 miles per hour. So far as the moon is concerned, the physical laws of space compounded with the problems of the fixed orbit of the moon make it extremely impractical as a military base from which weapons could be launched. In fact, not a single respectable military authority has yet explained the military potential of the moon. Its potential as a base for communications or as an observatory, however, are very important. For an astronomer, observations from the moon would make obsolete all the great observatories on earth; for from the moon, a view of the stars through a telescope would be unobscured by the curtain of the earth's atmosphere. What he would observe of the stars and of earth from the moon would be of tremendous scientific value.

It is true that rockets which can launch satellites into earth orbit or to the moon can also launch bombs across long distances on the surface of the earth; but using rockets to race for the moon may keep us from using rockets to shoot at each other. As long as men are going to fight among themselves, there is always a danger that a way may be found to use a peaceful instrument as a weapon—whether it be a rolling pin or a fence pole, a laser beam that can pierce a diamond or a rocket that can launch a satellite. I believe, however, that guided by God's law and their own intelligence, ultimately men will make proper and constructive use of the knowledge and power gained through the centuries with the help of God.

The accent on space today is peaceful. As a scientific venture, it has already proved of great value in knowledge gained, and promises more for tomorrow. As an economic venture, it is amply justified. The savings in life and property from the weather satellite alone will, I am certain, be worth all the money put into space exploration.

Through science, man's future in space and on earth can be

as bright and high as the stars. Science, handmaiden to theology, is the God-given key to the wonders of the universe whether we look down into the atom or up into the heavens. If we use it with trust in His commandments, all that we will do will lead to good for ourselves and others; for it will lead us to God. This is our great promise. Our destiny lies with Him, and the road to that destiny with us.

RELIGIOUS RESPONSIBILITY IN THE SPACE AGE

James A. Pike

Since the beginning of time, man has looked upward at the heavens and wondered at the mystery of the stars and the vast regions that encompass them. When Sputnik I orbited the earth in 1957, the stars and the space surrounding them came within our reach; and among us were those who expressed fear that man, so long earthbound, might in his exploration of space be moving against God's will.

This fear and doubt I do not share. As the poet Robert Browning said, "Ah, but a man's reach should exceed his grasp, Or what's a heaven for." We are charged by our Judaic-Christian heritage to serve God with all our might—with all the powers of mind, spirit, and body He has given us. Fashioned in His image, we have been given the gift of reason so that we may understand the world about us and our relation to it.

In the full meaning of this imperative, the exploration of outer space is morally justified, but only if we are made conscious of our greater religious responsibility. For while it is a duty to explore outward, it is even more urgent that, at the same time, we explore inward. For only by self-knowledge are we ever to eliminate the selfish motivation that has been destructive of our best and altruistic purposes since creation.

The discovery of new outer worlds has always had more appeal for mankind than self-discovery. Yet without the understanding and conquest of self that must follow, we cannot serve God's greater purpose. This is the great moral responsibility of the Space Age as, indeed, it has been through all the ages and times in which man has lived. The responsibility has not changed; only our awareness and sense of urgency are new and profound. The mind of man has released awesome energies and power; it has created the marvelous machinery for space exploration and atomic development. He must make every effort to adapt their use for service to God. Failure would mean surrender to evil and possibly total annihilation.

Space science, like science generally, is morally ambivalent. Science does not dictate our moral premises but may test our faith by confronting us with the necessity to make decisions concerning its implementation. Our rockets can carry bombs or instruments of exploration. We can transmit and receive messages millions of miles in space and to every corner of the earth, but the content of the message is all-important. Shall it be hate and mistrust or brotherhood and love? Science has created one physical world, but only our morality will make its inhabitants neighbors of one another.

Church membership in this country has risen from 15 per cent of our population in 1865 to nearly 70 per cent today. Mass communications are enabling religious programs and messages to reach into virtually every American home. But we still are a long way from accepting the message of brotherhood and love. Obviously, just being able to speak the message louder and to more people does not induce understanding and acceptance. To achieve acceptance of love and brotherhood by the minds and hearts of men, we must speak in the voice of our time, employing the techniques now at hand to probe the problems of individuals; for many of the things we think of as bad in man and once called "sinful," we now understand are due to psychological compulsions and conditioning. In the complexity of today's society, we cannot overcome this "sin" by making holy noises suitable for a past age. Counseling rather than preaching, treatment rather than penance are the more useful remedial procedures.

Science has penetrated every aspect of our life and so must religion. Religion must enter the marketplace. It must enable us to recognize that we are under the claim of God during our business hours as well as during those hours we pray to Him in church. We are to do His bidding with no time off. We must convey to our laity that, though individual piety is important, they and we must strive for a conversion to a higher piety, the development of a moral order in society that will bring us to an acceptance of God above all loyalties, whether they be national, racial, cultural, or ideological. Such global organization of the mind and spirit is our religious responsibility and sacred obligation for this new age, the Space Age.

Space exploration can be a source of inspiration. The wonders it is revealing of the universe magnify our vision of God. It increases in us the sense of awe and wonder that is very close to the heart of true religion and gives us a fuller appreciation of the greatness He has given us. This will enable us even to accept the discovery of other intelligent life elsewhere as a new revelation of His wonder and divine power. For in space and on earth, God reveals Himself. We need not search; we have only to look for His manifestation. For we are in touch with the true God, under claim to Him and subject to His judgment and grace no matter what our explorers find on other planets or in other worlds. This truth is at once our responsibility and blessing, our burden and salvation.

CONFLICT IN THE RACE FOR SPACE

Lillian Levy

Our substantial national effort in space is, in part, a response to the persistent differences—political and military—between East and West; and to that extent, conflict has accelerated our progress in the race for space. But our space program has also been marked by other conflicts—national rather than international in origin—which have had less positive consequences. These internal controversies are concerned with many aspects of our exploration of space. They have posed serious questions about such matters as the civilian mission in space exploration and military defense; the high cost and priority of manned lunar exploration and the economic effects of such vast concentrations of scientific and economic resources, including manpower, in space research and development. Even the standards established for selecting astronauts are not free from dissent. For the most part, the antagonists who have raised these issues are either on the sidelines of the space program or, if active participants, view it from a particular perspective. Their dissatisfaction or point of conflict may reflect self-interest, but such motivation is not necessarily inconsistent with either truth or merit.

No controversy or conflict would be complete today without the issue of woman's role and status, and space is no exception. This particular issue has been initiated by several well-known American women aviation pilots who seek to qualify as space pilots. They have complained that the space administration's rule for the selection of its astronauts—all male to date—unfairly discriminates against women because of the mandatory requirement that all applicants must be experienced jet test pilots. Authorities at NASA insist this is a necessary requirement and deny any discriminatory intent, but there are no plans for including women in the astronaut program.

Other reasons offered for excluding women are that the physiological differences in the sexes would pose special problems, such as fitting space suits, and this as well as the allegedly special research and training that would be required would add to the expense of the program. Their major value would be as biological specimens; but since data on the female organism in the space environment has been gained from observing female chimpanzees and other simians, there is doubt whether the data to be gained from observing woman would justify the effort.

No one, certainly none of the women pilots, would deny that women are physiologically different from men, and that they would require special attention and training. However, the would-be astronettes seriously question whether experience in jet test flying is essential. The women claim that any good pilot, with or without jet experience, can be trained to fly a spacecraft; and there is in fact substantial evidence that the need for extensive flying experience in high performance aircraft is highly exaggerated. Instrument piloting requires as much, if not more, skill and precision; a commercial pilot making an instrument landing, for example, is allowed less than half the margin of error permitted an astronaut in the crucial re-entry attitudes. The fact that none of the Russian cosmonauts had extensive experience as jet flyers is also significant. With modest conviction, the women flyers further suggest that observing women, as biological specimens, will surely yield data of greater interest and more beneficial to man than that gained from observing female simians.

The cause of the women received its best and most unexpected support from the launch on June 16, 1963, of Valentina V. Tereshkova of the Soviet Union. The world's first woman in space was a former textile worker and amateur parachutist. She volunteered and was accepted for cosmonaut training in 1961. She had no previous training as a pilot, and her subsequent piloting experience was limited to conventional aircraft only. During nearly 72 hours of orbital flight in Vostok VI, she handled her spacecraft expertly, and skillfully performed the re-entry maneuver and landing on June 19, 1963. Her space debut was part of a Soviet experiment in which two manned vehicles were launched into space for separate orbiting at the same time. Her orbital flight was almost twice as long as that of any of our jet-trained astronauts to date. Our American women flyers, all skilled pilots, hoped that the dramatic triumph of the Soviet spacewoman might advance their cause. But apparently the preeminence of Soviet women in the race for space was not viewed competitively by our government, and the American rules for space travel, so far as women are concerned, remain unchanged.

The orbit of the Soviet woman did, however, have significant influence on an entirely different controversy. It provided an incentive to our scientists, who have long wanted an opportunity to journey into space, to intensify their demands. The scientific community warned that the value of a manned lunar expedition would be reduced to an achievement for technology, not science, for machines, not men, if a scientist were not included in the three-man Apollo crew. But the requirement for jet-test flying experience, protested by the women, also proved a bar to most of our competent scientists. Supported by the Space Science Board of the National Academy of Sciences, the scientists took the position that it is much easier to train a good scientist as an astronaut than it is to train a pilot-astronaut to be a first-rate geologist or astronomer.

The successful flight of the Soviet spacewoman gave persuasive support to their cause. The scientists won their point, and in the late spring of 1964, NASA announced that it would seek forty physically fit scientists for the lunar venture. This was far less than the 500 recommended by the Space Science Board. They will train on a part-time basis only at NASA's Manned Spacecraft Center in Houston. In 1967 or 1968, five or ten of these astronaut reservists will be selected to be the first scientist-astronauts, but will not be on the first three flights to the moon. Flying experience will not be as much as that required of the other astronauts, and it may not include jet flying. Age requirements will also be less stringent, with less emphasis on youth and more on knowledge and scientific competence. Who knows, science may yet serve the cause of women. Perhaps a woman scientist may be among the applicants chosen.

In point of time, the oldest of the internal conflicts is that concerned with the civilian mission in space exploration and military defense. The distinction between these two aspects of our space program was challenged in the early days of the Space Age, and those who posed this challenge took the position that our national space activities could be pursued more effectively and efficiently as a unified program under either civilian or military direction. However, among those who supported the division, there was and still is conflict over which should be pre-eminent.

To some degree, all these issues have their roots in the apprehension as well as the challenge which were inspired by the Soviet launching of Sputnik I on October 4, 1957. Public fears were somewhat allayed by the reaction of General Dwight D. Eisenhower, then our President, who discounted the suggestion that this achievement was a threat to our national sceurity. He said, "So far as the satellite itself is concerned that does not raise my apprehensions, not one iota."

President Eisenhower also emphasized that space activities and military defense are quite distinct. On October 9, 1957, he congratulated the Soviet scientists on their space achievement and in a reference to our own Vanguard satellite program pointedly underscored its separation from the military work on ballistic missiles. This viewpoint was amplified on April 2, 1958, when he proposed to Congress that legislation be enacted to establish a separate agency under civilian direction for the scientific exploration of space.

This proposal and its peaceful accents were supported by Lyndon B. Johnson, then Senate majority leader. In his opening statement on May 6, 1958, at hearings before the Senate Committee on Space and Astronautics on the legislation which he introduced, Senator Johnson said, "... On all sides, there is wide agreement that ... the ultimate opportunity of space is not that of a final battleground. Freemen have no intention of rattling sabers among the stars ..."

In July, Congress enacted and President Eisenhower signed into law the National Aeronautics and Space Act of 1958, which opens with the declaration that "It is the policy of the United States that activities in space should be devoted to peaceful purposes for the benefit of mankind." Under the Act, NASA was established and assigned the responsibility "to plan, direct and conduct aeronautical and space activities." NASA's role and function are the dominant themes of the statute.

The military aspects of space exploration were assigned a residual or correlative role, separate and apart from the civilian or purely scientific objectives of the program, and were to be undertaken within the Department of Defense. A Civilian-Military Liaison Committee was to coordinate the military phases of space technology, such as launch and other such support, with NASA's nonmilitary undertakings. The President, assisted by the advisory services of a National Aeronautics and Space Council, was responsible for securing effective cooperation between NASA and the Department of Defense. In the event of conflict, the President was to be the final arbiter. The law also provided that the President, subject to review by Congress, could transfer from the Department of Defense projects related to the civilian program which it had previously undertaken.

The emerging NASA program necessarily included extensive transfers of projects from the Department of Defense. The first was the Vanguard, followed closely by Projects Tiros and Centaur and others. The best known was the transfer to NASA of the Development Operations Division of the Army Ballistic Missile Agency from the Redstone Arsenal, Huntsville, Alabama, together with the Division's personnel and facilities. The Division, since named the George C. Marshall Space Flight Center, is responsible for the development of Saturn, the giant super-rocket engine designed to carry three men in the Apollo spaceship to the moon in this decade.

As might be expected, this transfer prompted objections from the Army missile men who felt the civilian space effort was being strengthened at the expense of military development. The formal provisions of the 1958 Space Act, particularly the transfer authority given the President, were far from satisfactory to those who urged unity in space and who preferred consolidation to cooperation. The transfer of the Redstone missile group to NASA, in their view, only supported their contention that space activities were really indivisible, since the technologies for both military and civilian programs are virtually the same.

Among the most vocal proponents of this viewpoint was General J. B. Medaris, U.S. Army (Ret.), former commanding general of the Army Ordnance Missile Command. In testimony before the Congress in February 1960, General Medaris said: "From a purely technical viewpoint, there is so little difference between civilian and military space programs that there is no justification for their division and resulting duplication. For example, in the area of powerplants, both programs are concerned with a reaction-type engine, liquid or solid, whose functioning requires rather sophisticated control. This is a fundamental characteristic of every vehicle, whether it be a short-range ballistic missile used by troops in the field, or a more ambitious vehicle used in an interplanetary probe." He urged a consolidated space program under the Department of Defense in order that our weapons development would not lag and our space effort could go forward.

General Medaris indicated that many of his military colleagues endorsed his position privately. However, most of those still on active duty limited their public expressions to underscoring the pioneer role of the Army in rocket development and the basic similarity in space technology, whether a mission be for a civilian effort or a military objective. General James M. Gavin (Ret.), former Deputy Chief, Office of Research and Development, also agreed ". . . that one cannot physically separate the nonmilitary and the military space activities." But he recommended a unified space effort under NASA as a practical way to avoid inter-service rivalry within the Defense establishment. The advocates of a unified space program were not limited to the military. Adopting the same reasoning, W.M. Holaday, former chairman of the now defunct NASA-DOD Civilian-Military Liaison Committee, urged a unified space program but also recommended civilian direction.

There is much truth in the asserted identity of technologies for any effort in space; but the difference lies in application and purpose. The same general technologies are employed in the construction of a destroyer and for a vessel designed for oceanographic research. They differ, however, in function and objective. In this same sense, the civilian and military missions in space are distinct and different. To regard outer space as a universe for scientific pursuits is one thing. The conception of outer space as a military proving ground and a possible combat theatre is quite another. Both the character and emphasis of research and exploration are invariably affected by these divergent conceptions.

To resolve this issue more explicitly, President Eisenhower recommended in his message to Congress in January, 1960, that the Space Act be amended to "clarify management responsibilities" so that the concept of "a single program embracing military as well as non-military activities in space should be eliminated." He declared, "In actual practice, a single civilian-military program does not exist and is, in fact, unattainable."

At a White House press conference on January 26, 1960, he brushed aside the suggestion that separation of functions in outer space would or could cause confusion. When asked to distinguish between space exploration and defense, he said, "These things are different. They are for different purposes. Now that does not mean . . . that . . . if the Defense Department can find some space activity that can contribute to its defense—well, quite naturally we'd explore it. But the difference between space activity as such and defense is really quite marked and not nearly as confused as it is for example between say Air Force and Navy and the Navy and the Army and the three of them put together."

The Act was amended in the fall of 1960. NASA was again assigned responsibility "for the exploration, scientific investigation and utilization of space for peaceful purposes." The basic policy of civilian pre-eminence in space exploration and its dedication to peaceful uses thus remained unaltered; but the separation between the civilian space effort and defense was more definitely established. The amended legislation more clearly specified that the provisions of the Space Act in no way reduced the authority of the Department of Defense to maintain our military strength, including such acivities in space "as may be necessary for the defense of the United States." This merely underscored existing authority. The amendments also modified the President's role as arbiter and revised the civilian-military liaison organization so that duplication of effort might more effectively be avoided and cooperation more easily achieved.

The legislation to amend the Space Act resolved the conflict over the separation of space activities, but it did not fully resolve the conflict on the matter of pre-eminence in space. Members of the military establishment, who accepted the position that the military role in space was, indeed, different, questioned whether it was adequate, particularly in respect to manned space flight. This was especially true of the Air Force.

Manned space flight has been an Air Force objective for many years, an ambition pre-dating NASA. In March 1958, General Bernard A. Schriever, at a conference at the Air Force Ballistic Missile Division in Los Angeles, California, announced that the Advanced Research Projects Agency of the Department of Defense had asked the Air Force to report on putting a "Man In Space Soonest" (MISS). He indicated that the Air Force had been given the job of manned orbital flight. With the establishment of NASA a few months later, the Air Force MISS missed out on further development. Apparently a clearly defined military mission for man in space was not established.

NASA's Project Mercury became our national manned space effort, with technical support from the Armed Services, particularly the Air Force. Mercury launches were prepared and executed at Air Force launching facilities at Cape Kennedy. The Atlas booster used for the Mercury launches was developed by the Air Force; and its stations were used for tracking and data collection. These and other supporting and auxiliary services are in full accord with the policies of the amended Space Act. They are essentially no different from the logistic and other support that our Navy and Air Force supply for our scientific research in the Antarctic, whose use for military purposes is forbidden by treaty. As in Mercury, the Air Force has a support responsibility for Project Gemini, NASA's two-man spacecraft scheduled to fly a manned mission before the end of 1964 or early in 1965. The Air Force's Titan II will be the launch vehicle for the Gemini missions. Air Force boosters account for about 80 per cent of the satellites the United States has orbited.

A listing of Air Force space projects alone indicates that the primacy of our peaceful mission in space has not closed this domain to defense requirements. The list includes the development of large solid-fuel boosters; the Titan III with its storable liquid-fueled engines and solid-fuel supplemental boosters; research on ion engines and other types of advanced propulsion; MIDAS, SAMOS and a variety of surveillance satellites; the Discoverer satellites, which have studied the effects of radiation on small animals, human tissue and other materials, and which have been recovered in the air after capsule re-entry into the atmosphere by means of a technique developed by the Air Force; electromagnetics and bioastronautics; space weapons and other studies ranging from space-flight dynamics to space-defense support. In December 1963, it was also given the authority and the appropriations for a manned space station, the Manned Orbiting Laboratory (MOL). Its space budget for fiscal year 1964 was well over one billion dollars, more than 2 per cent the total defense budget.

The Air Force is firmly persuaded of the need for manned military missions in space and wants the authority to conduct its own program in order to demonstrate the military role for man in space—if there is one. What this mission may be, when all is said and done, is largely equipment testing, inspection, maintenance and repair, as indicated by General Schriever's own brief description of the MOL program in an address before the National Space Club in Washington, D.C., on May 20, 1964.

He said: "The MOL will be made up of a modified Gemini capsule and a pressurized module which will serve as a laboratory. It will be large enough to allow two astronauts to move around and operate equipment for periods of up to thirty days without wearing special space suits. In this shirt-sleeve environment, the astronauts will be able to serve both as test pilots and scientific experimenters in space. They will be able to test and evaluate experimental equipment and determine man's ability to use the equipment in the discrimination, evaluation, filtering, and disposal of data. These would be some of the functions required for possible manned space missions, such as earth observations, inspection, maintenance, and repair."

Gemini missions for two or three weeks and longer are also planned, as well as the development of a shirt-sleeve environment and studies aimed at permitting astronauts to venture outside the Gemini capsule for maintenance and repair work in outer space. Just what may be learned from MOL that will not have been discovered from NASA's Gemini and Apollo is not specified or indicated. General Schriever, in his address to the National Space Club audience, merely stated: "One of our most important tasks is to determine the military usefulness of man in space. This is the purpose of the Air Force Manned Orbiting Laboratory."

While the Air Force has voiced full and enthusiastic approval of the division of authority between our civilian space program and military defense, its leaders from time to time suggest that the strong accent on peaceful exploration of space under a civilian agency is incompatible with as complete a military posture as they believe to be necessary for our security in outer space. An Air Force general has suggested that the orbit of the Soviet's cosmonette, who had little special training in jet flight, may be evidence that the Russian spacecrafts are designed for the greater maneuverability which is essential for a space weapons system but which our NASA spacecraft, such as Mercury, Gemini, and Apollo, do not possess. Air Force Chief of Staff, General Curtis E. LeMay, as well as other of the service's top generals, remind the public of the threat of a bomb-carrying orbiting satellite voiced by Soviet Premier Khrushchev and other Soviet officials. At the same time, it is readily admitted that it is much easier to hit a target on earth with a ground-based missile than to hit the same target from a satellite moving with a velocity of 17,500 miles per hour in space.

This conflict over man in space, of which the Air Force is the author, is not yet resolved; but it should be recognized and re-emphasized that the quarrel is not with NASA. NASA's manned space missions are no bar to the Air Force if it can demonstrate a valid military mission for a larger role in this field. Nor is it a matter of competition between NASA and the Air Force for authority and funds. In fact, the conflict is really between the Air Force and the top leadership in the Department of Defense. As the Space Act makes clear, the Department of Defense has the authority, subject only to necessary budgetary approval, to enlarge the Air Force role in space which, quite apart from its support for NASA, already is quite extensive. The Department of Defense, with the full support of

President Johnson, has maintained the position that no additional defense funds will be allocated for space in the absence of a clearly demonstrable "military requirement."

Conflict as well as many ambivalent stresses are also apparent in our budgetary policies for space. NASA's appropriations for its first year of operation were a relatively modest $338.9 million. Subsequent budget requests and appropriations rose sharply, and in January 1961, President Eisenhower's request for $1.1 billion for NASA for the fiscal year 1962 was granted by the Congress. Shortly after assuming office, President John F. Kennedy asked NASA to re-examine President Eisenhower's January budget. NASA responded with a request for more funds, and in March, the NASA budget was amended with an increase of 12 per cent of the $1.1 billion. When about a month later the Soviets achieved another space triumph with the orbit and recovery of Yuri Gagarin, the first human to travel in space, we were faced with the fact that the United States had again placed second best in space.

On May 25, 1961, President Kennedy, in a State of the Union message before a joint session of the Congress, declared that the United States intended "a clearly leading role in space achievement." He called for an accelerated effort in space communications and meteorology and for a national commitment to the goal "before this decade is out, of landing a man on the moon and returning him safely to the earth." He warned that this would require a large investment of our resources, both in manpower and in money. He estimated a budgetary increase totaling from $7 to $9 billion in the next five years; and he asked for an additional half-billion dollars for the fiscal year 1962.

The NASA budget for the fiscal year 1962 was again amended in May 1961. Together with the March increase, a total of $1.8 was appropriated, representing an increase of 61 per cent for the year. Project Apollo, NASA's program for manned lunar exploration, became a national goal of the highest priority. Its public acceptance was reflected in the approval by the Congress of about $3.6 billion for the fiscal year 1963, an amount almost twice that allocated for space in the previous year. But for fiscal year 1964, NASA's request for $5.7 billion was reduced by one-half billion. For fiscal year 1965, Congress again balked, cutting NASA's request for $5.3 billion by about $200 million.

These legislative reductions in our space budget represent neither a revision in our space goals nor their rejection. They do, however, reflect a mixture of concerns—political, economic and scientific—and a growing uncertainty about the rationale of our space program, particularly lunar exploration by man. The larger political as well as economic considerations were accentuated when President Kennedy announced a manned lunar landing in the 1960's as a national goal. Politically, it appeared that he was extending the cold war into outer space; and this has, to a large degree, been accepted as his intent, although from his later actions and statements, it appears that President Kennedy had a much different purpose in mind.

Far from intensifying East-West rivalry, it appears that his proposal was designed to secure cooperation. He hoped and expected that, as our capability in space progressed and was clearly demonstrated, our offers of cooperation would more readily receive an encouraging response. This expectation largely motivated his offer to the Russians of joint exploration of the moon, which he made in an address before the United Nations in September 1963.

The national reaction to this proposal fell far short of wholehearted support. It was viewed by some as simply a political stratagem to demonstrate that the Soviet Union and not the United States was dragging its feet on cooperation and, perhaps, as second thoughts by the Administration on the urgency of reaching for the moon. Some NASA officials went so far in opposing his proposal as to state that such cooperation was technically unattainable or, at best, impractical and difficult. However, James E. Webb, NASA's Administrator, pointed out several areas in which cooperation might be effected.

President Kennedy found little public sympathy for his offer to change the race to the moon to a team effort, and the wisdom and purpose of his proposal were questioned in the Congress. On September 23, 1963, President Kennedy clarified what was behind his offer in a letter to Representative Albert Thomas of Texas, in which he wrote: "This great national effort and this steadily stated readiness to cooperate with others are not in conflict. They are mutually supporting elements of a single policy. We do not make our space effort with the narrow purpose of national aggrandizement. We make it so that the United States may have a leading and honorable role in mankind's peaceful conquest of space. It is this great effort which permits us now to offer increased cooperation with no suspicion anywhere that we speak from weakness."

In spite of this explanation, the public and Congress continued to view our goal for a manned lunar landing in the 1960's with mixed attitudes. The legislative reductions in space spending reflect countervailing pressures for economy but, paradoxically, without curtailment of our national goals. Although at the hearings on appropriations most of the criticism was directed to our manned lunar venture, comparatively little has been cut from the billions allocated for the manned space program. The reductions were proportionately much greater in the appropriations for the space sciences, advanced research and technology. Apparently the desire of the Congress to economize on space spending was counterbalanced, if not outweighed, by a fear that a substantial reduction in the lunar program might turn out to be politically hazardous if the Soviets should succeed in a manned mission and thus demonstrate that they were pushing to the moon faster and more effectively than we are. In terms of our world stature and political realities, a withdrawal or a substantial reduction of our lunar venture appears almost as difficult as a retreat in Laos or Vietnam.

In the past, the launch of a Soviet manned satellite served as a powerful deterrent to any reduction in space spending. It was a reminder of their superior rocket power which made it possible for them to orbit and recover massive manned and unmanned payloads of five tons or more—bigger if not better than the payloads we could then put in orbit. The Russians have had this capability since 1961, while our best estimates are that not until 1965–1967 will our Saturn rocket enable us to equal the 1961 capability of the Russians and even move ahead. But this still appeared a long way off when the Russians again demonstrated their superiority in manned space flight with their dual launches in June 1963 of the Soviet cosmonette and her male colleague, who preceded her and remained in orbit more than five days. Yet, this Soviet feat did not, oddly enough, inspire or convey to Congress the sense of urgency that had followed all previous Soviet manned spectaculars. It came at a time when the glamour and high excitement that characterized the earlier days of the Space Age were beginning to fade. The manned Mercury flights had come to an end, and the prospect of more advanced manned excursions was more than a year away. The lustre had even dimmed on the bright promise of a lunar landing. This was still a national objective, but enthusiasm had given way to appraisal.

Whether it was wise officially to declare our manned lunar program a major front line in the East-West conflict is a judgment that history will ultimately render. Carl Dreher, who is both a competent engineer and a respected science writer, has said: "It is immoral to put a program of this magnitude, by far the greatest ever undertaken by man, on the level of a teenager's drag race. Worse, it is stupid, for the chances of loss of national prestige are greater than the possible gain . . ."

Among the scientist-critics of the lunar program are twenty-four of America's fifty-five living Nobelists. They are also largely critical of what they consider a "crash" approach which may hinder rather than advance scientific objectives. At a meeting in May 1963, at Gustavus Adolphus College in St. Peter, Minnesota, there was general agreement with the recommendation by Dr. Peter J.W. Debye, Nobelist in chemistry, now at Cornell, that a more temperate pace be undertaken and that the accent should be on instruments rather than man in a program for lunar exploration aimed at scientific objectives. There was also opposition expressed by some of the Nobelists, notably Dr. Linus Pauling, winner of a prize in chemistry and the Nobel Peace Award, and Dr. Polykarp Kusch, physicist at Columbia University, to the billions invested in space, when more pressing earthly needs remain unsatisfied.

Others, scientists and engineers, were also criticizing various aspects of the space effort, and in June 1963 the Senate Committee on Aeronautical and Space Sciences invited the testimony of several of our leading scientists "to assist the committee in evaluating (a) the over-all goals of our space exploration effort in comparison with scientific aspects of other national goals, and (b) the relative emphasis on the various projects within the space program in connection with its consideration of NASA's authorization request for fiscal year 1964."

Among those called to testify was Dr. Lloyd V. Berkner, past chairman of the Space Sciences Board of the National Academy of Sciences, who said that the race for space between the United States and the Soviet Union was inevitable. He said: "Men everywhere see, in the conquest of space, the peaceful demonstration of the superiority of one of the two competing systems of economic organization—capitalism versus communism. The conquest of space has become a symbol of the challenge of each system to demonstrate its superiority . . ." In his judgment the scientific objectives were an integral part of the political objectives in the space race; and, he added, ". . . The implications of allowing our technology to fall to second-rate stature, with respect to space, are less a matter of personal pride and more a matter of technological posture in a 'cold war'—a posture recognizable on all sides. This posture involves 'prestige' only in the sense that genuine military strength involves prestige."

Many of our leading scientists who support the manned lunar program have said that it cannot be justified on the basis of scientific contributions alone. Dr. Martin Schwarzchild, astronomer at Princeton University, expressed this viewpoint quite frankly at the Senate hearings, stating: "I think it would be plumb dishonest for me to maintain that I think that what we can learn on the moon, however fascinating it will be for the understanding of the origin of our planetary system, can be worth by itself the enormous effort that it will take." The job of gathering scientific data from the moon can be done better and far more cheaply by unmanned instrumented lunar probes. NASA spokesmen agree quite candidly; but they also take the view expressed by Dr. Colin S. Pittendrigh of the Princeton University Department of Biology who said that "we still cannot build a robot scientist." The biologist added that since such landings ultimately will be made, the manned program should be supported; but it should be preceded by purely instrumented exploration.

While most scientists would agree with Dr. Harold C. Urey, Nobelist in chemistry, now at the University of California in La Jolla, that "the most durable results of the program are almost certainly the scientific facts which will be learned and studied," the proponents, including Dr. Urey, justify manned lunar exploration for the broad advances it provides in several directions, not just in politics or science. It derives its importance as a stimulus to national energy as well as from its contributions to technology, the expansion of industrial, educational, and research facilities, the increased demand for excellence in training and education, and renewed pride in intellectual effort. According to Dr. Schwarzschild, our efforts in space have shaken us from the cult of the average that seemed to have set in after World War II, and even such intangible incentives are "overwhelmingly important."

The majority of scientists who support the manned lunar program will agree that it is justified by its broader benefits, many of them nonscientific; but this, apparently, is the limit of their unanimity as far as space research is concerned. One would like to believe the assertion by Dr. Lee DuBridge, President of the California Institute of Technology, that if scientific research was our only objective in space, "competent scientists could come to an agreement on what that program should be like"; but the evidence is otherwise. Each scientist who has testified before the Congress on the conduct of our space program has wanted more emphasis on his particular discipline; even among those in the same field, there are differences. Dr. Pittendrigh of Princeton and Dr. Joshua Lederberg of Stanford University Medical School would like to see greater effort directed toward exobioloby, or the search for extraterrestrial life. They would like this to have the same priority as manned lunar exploration; and, according to Dr. Lederberg, such research "would, by itself, justify the cost of the space program." However, such research is opposed by Dr. Barry Commoner, biologist at Washington University in St. Louis.

The most virulent and extreme opposition offered by a scientist to the space effort was presented to the Senate Committee at the June, 1963, hearings by Dr. Philip Abelson, editor of *Science* magazine, the highly respected and influential publication of the American Association for the Advancement of Science. Dr. Abelson, who is also director of the Geophysical Laboratory of the Carnegie Institution of Washington, D.C., charged, among other things, that NASA's program "is having and will have direct and indirect damaging effects on almost every area of science, technology and . . . may delay the conquest of cancer and mental illness." Our civilian space effort, he said, is adversely affecting the technical resources of the weapons laboratory and thus is a threat to national security. According to Dr. Abelson, scientists working in weapons development "feel keenly public antipathy for them and their work. They operate under necessity for complete secrecy at all times . . ." motivated by patriotism. Having discovered that space research rates high as a patriotic endeavor and offers respectability as well as status, the weapons scientists, according to Dr. Abelson, are rushing into the space program. He also accused NASA of raiding industry and pressuring students and scientists in universities to get personnel for its program. In spite of all these charges, Dr. Abelson said he would not discontinue the lunar program. He objected, however, to its priority as a national goal.

Dr. Abelson failed to present any basis for his rather far-fetched idea that NASA is staffed largely by scientist refugees from the sinister and security-ridden weapons laboratories. Indeed, he has yet to document any of his allegations. As an alternative to the documentation he lacked, he suggested that Congress look into the matter. He also insisted at the hearings and, again, more recently in an editorial in *Science*, 24 July, 1964, that those scientists who disagree with his views are motivated by self-interest and the desire for federal support for their research. He also said that fear was behind the failure of other scientists to be more critical of space spending, that the uncritical scientists are afraid of "incurring the enmity of powerful foes," and that "prudence seems to dictate silence."

His colleagues at the Senate hearings took issue with many of Dr. Abelson's unsupported accusations, as others have since. Dr. Pittendrigh pointed out that there is no lack of funds for cancer research. He also questioned whether funds spent on space would automatically or easily go to the other purposes. He said: "The lack of support for many of them is because it is politically impossible to get the money to them"; and if money were withheld from space, it would not more easily go to other services.

Both Dr. DuBridge and Dr. Simon Ramo, scientist and engineer and vice chairman of the board of directors of Thompson Ramo Wooldrige, Inc., challenged Dr. Abelson's allegation that scientists are being drawn away from other scientific projects by the attractions of space research. Dr. DuBridge said that most scientists go into the fields in which they feel they can do the most fruitful work; and Dr. Ramo pointed out that if the "program uses up talent, it also produces talent." He said there was, at present, no shortage of scientists and engineers, and that "in fact . . . we have today an overexpanded capability . . . geared to handle more projects than the nation has in the offing." He also denied that there was any evidence, either in his industry or others, that NASA was raiding and robbing them of their scientific talent. Dr. Ramo observed that government is not economically competitive with industry in salaries for top-notch people.

Dr. Ramo's statements are borne out by a report issued July 12, 1964, by the National Academy of Sciences based on a two-year study of the uses of scientific and engineering talent in the United States. The study showed that scientists in industry make as much as 25 to 50 per cent more than their peers in government and that only about five per cent of the nation's 1.7 million persons engaged in scientific and engineering work are employed in the space effort. "Raising salaries is only one of several measures that must be taken if the government is to attract and retain its fair share of the nation's best scientific and engineering talent," the report stated. While there are no shortages of scientists and engineers, they occur for the most part in the large scientific and technological undertakings because of a lag in training in the new technologies which are required in such massive programs. In other areas, there are the surpluses in manpower that Dr. Ramo mentioned.

The public and Congress continue to have alternating thoughts about our manned lunar program. There is a growing inclination to let the moon trip slide into the next decade; but this is not in accord with our desire to place first in the conquest of space. While we have achieved pre-eminence in meteorology, space communications, and other unmanned scientific space projects, the venture of man himself—his landing on the moon—still remains the winning symbol. The minor trimming of the NASA budgets for fiscal years 1964 and 1965 is evidence that we are not yet willing to concede first place to the Russians in this phase of space exploration. But fiscal, as distinguished from political or scientific realities and necessities, are not quite so flexible or indecisive.

When Congress considered and approved President Kennedy's recommendation that a manned lunar landing in this decade be supported as a national goal, it was informed by NASA that the over-all cost for such a project would be $20 billion to $40 billion. With the assurance of Congressional support, the space agency planned accordingly for this vast effort, contracting for all the essentials it could foresee.

Almost all of our space programs are, of necessity, projections to a goal or what is termed "mission-oriented." The goal or mission must first be set and approved before any plans for its achievement can be undertaken. Once having decided on a goal, the target date for reaching it, while not mandatory, is crucial. It is the focus for intelligent program planning and management; and in considering appropriations for the goal, the over-all cost of the program should serve as the guide for each annual budget.

In a continuing development effort such as Project Apollo or its forerunner Project Gemini, a reduction in funds—even a small percentage—may hamper the effort and retard it; and the savings would only prove illusory. A reduction in the number of flights will not necessarily cut down the cost base either in Project Gemini or Project Apollo. As Mr. Webb has pointed out to the House Committee on Appropriations, which was considering the NASA budget for fiscal year 1965, the same support and manpower will be needed if we fly once in three months or once in six months. The operating burden accumulates in almost direct proportion to the time required for the completion of the program. If the pace of the lunar effort should be slowed down by a matter of years because of budget reductions, the operating burden "absorbs an exorbitant share of the total expenditure," Mr. Webb emphasized. There is no savings in a stop-and-go approach to space, he said. Dr. Ramo also spoke to this point at the Senate hearings, stating: "We should look at the possibilities, make our choices, then be decisive in implementation, eliminating on-and-off nervousness."

In space, as on earth, nothing succeeds like success, as the outstanding achievement of Ranger 7 on July 31, 1964, has illustrated. The major aim of the probe was to photograph the lunar surface adequately so that a safe landing site could be selected for a future manned lunar landing. The cameras on Ranger 7 took and transmitted over 4,000 photographs of the moon. The probe has been hailed by scientists here and abroad as a great triumph. In praising the work of NASA scientists who worked on the Ranger project, President Johnson indicated that any move to withhold support for the manned lunar program would be "penny-wise and pound-foolish." It appears certain that supplemental funds for the lunar program which were withheld for fiscal year 1964 are now more likely to be appropriated. The flow of funds for space will probably continue unimpeded, at least for the next twelve months.

Another of the controversies on our space program is that concerned with its impact on our economy. It has sometimes been said that our space spending is done at the expense of the development of much-needed new commercial markets. Those who hold this view call attention to the tremendous advances in commercial markets in western Europe, particularly Germany, where unemployment is virtually nonexistent. It is not just dollars that are being diverted from consumer production, according to these critics, but vital scientific and technological manpower whose ingenuity and talents are required to create new commercial products.

The fact is, however, that most of the creative scientists and engineers now in space research have come from universities or research institutions. Few, if any, have come from industries concerned with commercial production. A careful review will reveal that space investment has had a salutary economic impact on the whole economy. In Utah, for example, missile-making projects have more than counterbalanced the general and continuing decline in farm employment, the near stagnation of the mining industry, and reductions in rail transportation jobs. Economic growth directly attributed to space and space-related industrial development has been reflected by a substantial increase in the sale of new cars and trucks (36 per cent in the first quarter of 1963), new construction in home and apartment buildings, and a five per cent increase in over-all retail sales. Aerospace programs have boosted employment, business activity, and investment in Long Island. According to one survey, employment in the Nassau-Suffolk area in Long Island rose by 7 per cent in 1962. Business showed a general 10 per cent increase over 1961. The Delta Gulf areas in Mississippi, Alabama, Louisiana, and Florida are others that have benefited substantially from space spending. As to the question of commercial development, a review of the benefits in various new products that resulted from the radar programs, atomic energy programs and the missile programs of the past few decades will demonstrate that the economy, in the long run, benefits substantially from investment in scientific programs that may not have immediate direct application to commerce.

As advances are made in disarmament, the orderly transition of our defense establishments into other fields of production becomes crucial. Space is now a permanent government activity; and it offers the best immediate outlets for the talent and machinery now used for defense. Realizing this, more and more of our missile-makers are turning to space production.

Perhaps the best argument for the economic value of space is that an important segment of western European industry has called for a vastly stepped-up space program to be projected over four years. In the summer of 1963, nine countries, represented by 125 companies and scientific groups, formed an organization called Eurospace. The countries are Britain, France, West Germany, Italy, Belgium, the Netherlands, Norway, Sweden, and Switzerland. The initial report of Eurospace said: "Space research and realizations are and will be an essential motor for economic expansion, bringing about in many fields an improvement in industrial techniques, and are thus at the origin of a great deal of progress." The organization report warned that if Europe did not get into the space race now it would find itself quickly left behind in the scientific, technical, and economic fields.

It seems that many of our difficulties and controversies stem from the fact that our space program is more than a scientific and engineering undertaking. Initiated, directed, and financed by the federal government, it is equally an integral part of national policy and purpose. Its success and failures are, therefore, judged not only in relation to advances and retardations in science and technology but also by the extent to which national commitments and goals are attained or thwarted.

These dual aspects of space research and exploration are by no means parallel or identical. In strictest terms there are no scientific failures; for so long as time is not of the essence and only devotion to learning is the predominant motive, failure and success are both royal roads to scientific truths. The history of science has traversed both these roads, with eminent credit to our understanding and achievement. It is only when scientific pursuits are enlisted in the service of national preeminence or international rivalry that failure is turned into frustration and success into national and international acclaim. It is these overriding characteristics of our space program that account for the alternating moods of proud exhilaration and of brooding pessimism and for many of the questions and doubts about the course of our race in space.

THE UNLIMITED FUTURE

John Paul Stapp

During the last Ice Age in western Europe, between 35,000 and 10,000 B.C., there lived men comparable in stature, physique, and intelligence to modern Europeans. Though their human potential was no less than that of modern man, the urgent business of survival in a climate too severe for agriculture left them neither time nor energy for the discoveries and inventions that lead to technical advancement. The artifacts they left in caves and campsites in southern France and the Pyrenees reveal a respectable artistic talent, but a technology limited to tools and weapons of flint, bone, antlers, and ivory which, in 100,000 generations, did not advance beyond minor improvements in craftsmanship and variety.

Only when the climate of Europe became more congenial was it possible for these men of the Stone and Ice Ages to begin the transition to historic civilization recorded in the past 6,000 years. As community life in agrarian villages developed, with organized division of labor and food storage, the prehistoric stone and bone phase of mankind was ended. The organized community, the domestication of animals, and the cultivation of crop plants and their storage to improve the logistics of survival characterize the second phase of man's technological progression.

The third phase was the flowering of human ingenuity in handicrafts and the conversion of mineral and plant resources of metals, ceramics, and textiles. The progress of the first five thousand years of recorded history was limited largely to what could be accomplished by human and animal muscle power directed by human intelligence, initiative, and acquisitiveness. Fortifications, castles, and cathedrals were no small monument to that power.

A little respite to men and domestic animals came from the invention of water wheels and windmills. But the millennium of technological progress began in the fourth phase, with the development of power resources and machines which emancipated mankind from the limitations of muscle power and water wheels. From creative minds came the individual and consecutive inventions of gunpowder and chemistry, steam engines and engineering, electricity and physics, microscopes and medicine, printing and libraries and universities. Not the least of these scientific and technological developments was the scientific method, evolved from trial and error, inductive logic, and systematic controlled experiments. The scientific method or attitude became an objective devotion to truth derived from factual data, rigorously excluding opinion and biased judgment.

In the last century, technical and scientific progress advanced beyond the contributions of the individual. Communication between scientists provided criticism; collaboration developed concepts and inventions; the formulation of a scientific language facilitated the accumulation of organized information and its dissemination to succeeding generations of scientists who could continue the pursuit of a problem from where their predecessors left off. These developments advanced mankind to the fifth phase, the current Revolution of Technology, which has brought in-line assembly manufacture, vast increase of electric power resources and applications of atomic energy, electronic communications, air conditioning and refrigeration, the conquest of disease by sanitation, immunization and antibiotics, jet air transports, and computers.

The technological progress of the past twenty-five years of this fifth phase appears infinite by comparison with twenty-five millennia of the European Stone Age, keeping in mind that both these eras had in common the same strain of Homo sapiens having the same percentile ranges of physical and intellectual capacities! However, the most important development is the institutional organization of research. The coordinated effort of specialists and institutions in all phases of a problem has multiplied the efficiency of investigation and contributed to the success of an undertaking of such magnitude as the Manhattan Project. The Manhattan Project, which enormously accelerated atomic fission and fusion research, established a pattern for the organization of the even more mammoth national research programs required for the exploration of space, such as Project Mercury.

Behind the headlines that hailed the success of each Project Mercury orbital flight was an organized effort actively involving more than 19,000 people with all degrees of training and skill deployed in sixteen ground stations scattered around the world, sailing the oceans in twenty-eight ships, and flying in more than two-score aircraft. Never in human history have so many people so widely separated worked together on a single scientific experiment. A revolution in scientific research and technological development, this highly organized systems approach has opened the sixth phase in man's development—the Space Age. The beginning of this great scientific and technological renaissance merits more detailed consideration, particularly in its projected lines of development and future implications for the human race.

The most important development in this phase is man's

conquest of the limitations so long imposed on him by environment. Today, man has the mastery not only of the temperature of his environment, so that he can enjoy the fruits of hydroponic gardens in the Antarctic, artificially lighted and heated by electricity from atomic generators, but he is also able to create and maintain closed environments for surviving and working comfortably under the sea in submarines, in the stratosphere in airliners, in orbital space in artificial satellites. This has been done by extracting, accumulating, and converting energy from the environment to provide power for maintaining and controlling atmospheric temperature, pressure and viable composition within his inclosure while excluding all adverse environmental factors.

Man will achieve his supreme triumph over the most extreme environmental hazards he has yet dared to challenge when he succeeds in establishing observatories and scientific bases on the moon. To minimize logistic support from earth, these lunar colonies will require self-sufficient life support in a balanced vivarium, where the gaseous, liquid and solid chemical cycles of earth-life are reproduced and maintained within a shell. This shell must exclude the high vacuum at the moon's surface, the extremes of heating and cooling during 330½ hours of day and 330½ hours of night under conditions of total heat and light radiation, radioactive emanations from solar storms, and all sizes of stray meteorites. Whatever materials can be found on the moon for constructing this shelter will by that much reduce the costly cargoes brought from earth. Power requirements can be met with electrical energy obtained from solar cells, transforming the lethal, unfiltered sunlight on the moon's surface from a formidable hazard into the indispensable resource for survival.

In essence, this will be a miniature earth environment, described by Dr. Hubertus Strughold as a "terrella" or miniature earth. In Dr. Strughold's definition, this is the smallest completely independent and self-perpetuating package of the elements in our world needed for sustaining human life. It compares with a balanced aquarium in which plant and animal life are mutually self-sufficient and self-perpetuating. The terrella must solve simultaneously all problems of human earthly existence. Man exists by having adapted himself through evolution to his terrestrial environment; man lives and will conquer the universe by using the forces of nature to overcome nature and adapt the environment to his requirements.

The scope of man's technical advancement has been limited only by the energy at his command. Up to the middle of the twentieth century this energy has come from combustion of vegetable and mineral hydrocarbons, hydroelectric generators, and electrolysis in metal plate batteries. Today, atomic fission provides power for submarines and for some public utility generators, although at a cost higher than for conventional electric power systems. Atomic power plants of a size, weight, and longevity adapted to space vehicles in journeys of up to three years' duration have been developed at a cost justified only for research.

A more ingenious development for the same application is the fuel cell, in which electrolysis is reversed to produce electricity by the controlled combination of hydrogen and oxygen. A mixture of two parts hydrogen and one part oxygen can be exploded by a spark to produce water. Combined much more slowly through porous electrolytic barriers, current can be drawn at voltage proportional to the pressures of combining gases. A variant of this is the biological fuel cell, in which the combining gases are by-products of bacterial metabolism. Here lies a possibility of incorporating a fuel cell into the sewage-disposal system of a space vehicle, and eventually into that of a lunar colony, so that an electrical energy by-product can be derived from sewage processing, at the same time producing pure drinking water without distillation. Such biological fuel cells could be economically feasible in city sewage-disposal plants, where the gases of activated sludge are sometimes burned for heating.

The most significant energy source whose exploitation has

been accelerated by space power requirements is the direct conversion of solar heat to electric energy by photocells and thermocouples. The very first space satellite put in orbit by the United States, the Vanguard I, no larger than a grapefruit, derived its power for electronic transmissions from transforming sunlight into electricity. Its weak signals clearly received by astonishingly sensitive ground stations transmitted data permitting a more accurate appraisal of the earth's shape, which appears to be slightly oblate and pearshaped rather than spheroidal. With no way of shutting this satellite off, it has continued to beep uselessly and may do so for another hundred years from inexhaustible solar cells and indestructible transistor circuitry.

A much more sophisticated solar energy power plant, with self-orienting vanes to gather maximum sun-power, was used in Mariner II, the missile that made the journey to intercept the planet Venus. Power enough to transmit many channels of data and to operate complex internal controls of this missile was produced during several months of flight, with successful reception at more than 40,000,000 miles from the earth. The unfiltered sunlight of space provides a very pure source for precise experiments in solar energy conversion to obtain data for a moon power-plant design. Such a power plant would be based on scientific feasibility; operational experience with a large-scale lunar power plant would provide performance data for designing an economically feasible power plant to generate electricity from sunlight at the earth's surface.

The vastness of the solar energy resources on the earth's surface quickens engineering imagination. The earth has a surface area of 510,101,000 square kilometers. Of this, 148,847,000 is land, of which 30,000,000 or approximately 20 per cent is desert. The deserts of the earth are eminently suited for solar-electric power-plant locations because both inhabitants and clouds are scarce. Each square kilometer receives 10,176,000 large calories of heat energy per minute, equivalent to 11,832.65 kilowatt hours per minute. In 720 minutes, or twelve hours of sunlight, this would amount to 8,533,872 kilowatt hours' equivalent per square kilometer per day. The sunlight on 30 million square kilometers would then be 255.5853 million million kilowatt hours. If one-tenth per cent of this energy could be collected and if a process having an efficiency of one per cent converted it to electricity, the product would still be over 2.5 billion kilowatt hours of electricity. A population of a little over five billion people would therefore be rationed to half a kilowatt per person per day, or about the power consumption of a middle-class American. Considering that photosynthesis by chlorophyll in plants stores solar energy in chemical change with an efficiency of 4 per cent, this is a conservative estimate for energy retrieved from sunlight.

Long after a population increase on earth has resulted in a power consumption that has burned up the last of the available hydrocarbon resources, the human race could still carry on with energy obtained from sunlight. One can imagine a power plant consisting of a cluster of large captive balloons covered with solar energy cells floating above the desert power station at a favorable altitude, feeding current through the mooring cable conductors for storage or for transmission to consumers.

Another source of heat energy that can be tapped is the heat of the earth's crust. At a depth of 15,000 feet, the temperature is 282° F., that of live steam at 100 pounds per square inch pressure. A very deep well could provide heat for a closed steam turbine to generate electricity by piping down water and bringing up steam under pressure. This already has been done on the slopes of volcanoes, where heat is near the surface. Heat wells for steam electric generators could be located in conjunction with desert solar electric power plants as a standby reserve system.

Unlimited low-cost electric power can provide solutions for many problems besetting the present and menacing the future of mankind. Air pollution by combustion products of hydrocarbon fuels can be eliminated when battery-powered electric automobiles replace gasoline and diesel vehicles. Parking meters with recharging plugs can replenish the batteries while the car is away from the home garage recharger. Electric air-conditioning will replace home furnaces. Refineries need no longer fill the air with volatile poisons of high-test gas production. Quiet electric power will reduce the noise level of city streets.

As human population increases, the water problem grows more acute. Cities, towns, and farms will have ever less option to locate near water resources. Water will be brought to them instead. At present, the city of Los Angeles finds it more economical to bring water more than two hundred miles across the desert by aqueducts from Lake Mead than to de-salt or distill potable water from the adjacent Pacific Ocean. Research on the problem of extracting fresh water from the ocean is no longer concerned with methods but with cost accounting. Bringing the power cost down to a competitive level with natural fresh water transported a great distance will make large-scale fresh-water extraction from sea water feasible. Deserts near the ocean can become irrigated farms, ranches, and forests by using de-salted ocean water.

The entire food production needs of a growing world population may not be met by putting more land into agriculture with such irrigation, nor even by increasing productivity with fertilizers. Direct synthesis of nutrients from inorganic resources offers a challenging solution now within practical reach. Plants can take nitrates and ammonia from the soil and synthesize amino acids which are assembled into proteins by enzyme catalysts, promoting chemical reactions selectively. Continuing research on the low-cost synthesis of amino acids and from them, edible proteins, is the first step in direct synthesis of food.

A more difficult challenge is the invention of methods to synthesize carbohydrates directly from carbon dioxide and water in competition with the photosynthesis of plants. A cheap, direct way of combining water and carbon dioxide molecules into edible sugars and starches may mean man's emancipation from hunger and a way out of the Malthusian dilemma. Dependence on crops vulnerable to weather and pestilence will no longer determine feast or famine for mankind. Through synthesis of proteins and carbohydrates, man can control the water, nitrogen, and carbon cycles of nature instead of being limited by them in the struggle for survival. This now appears within reach.

Research to provide adequate and palatable synthetic diets for astronauts who will travel into space for months and perhaps years has had significant success. The diets contain essential amino acids, vitamins, the requisite salts, glucose as a source of carbohydrate and ethyl linoleate a a source of essential fat. One cubic foot of synthetic foodstuffs provides a month's supply of 2,000 calories daily and all required nutrients. This does not mean that synthetic foodstuffs will replace natural foods, particularly the proteins, but that they will supplement and augment the natural supply. Products less palatable to man can be used to feed livestock, poultry, and fish.

As a source of food and minerals, the ocean is scarcely explored and barely exploited. Much has been said about plankton, the tiny, single-celled floating plant life, which is abundant enough to be the basic diet of some species of whales. A way of harvesting plankton rather than hunting the limited supply of whales could create a large new source of edible oils and proteins. With cheap electric power, the reclamation of minerals from ocean water, such as magnesium, could be expanded. The floor of the ocean is only beginning to be thoroughly mapped, and eventually it will be gleaned of sedimentary ores such as manganese which is found in lumps on the bottom of the Pacific. Submarines and gigantic diving bells over mine shafts may facilitate the next revolution in offshore mining techniques.

The mass logistics of transportation between continents will bring about two very important developments that will make use of atomic power plants. With atomic power, the size of submarines is no longer power-limited. In the tranquil depths beneath surface ocean storms, submarines with the cargo capacity of today's largest freighters will move safely on schedule at several times the speed of their surface predecessors. Behind them they may tow submerged trains of plastic sausages enclosing bulk cargoes of liquid or gel chemicals such as fertilizers, solvents, paints, lubricants, wine, molasses and petroleum, which could be easily detached for unloading, with increased turn-around speed for the submarine-tractor.

Above the ocean, great flying ships of a size capable of landing only on water can carry cargoes of a higher transport speed priority. The weight of shielding of an atomic power plant is taken care of by buoyant displacement in submarine and surface ship power plants. For flying ships, the solution is one of relative sizes. A 5,000 ton flying ship can be propelled by a 50-ton atomic power plant, which would be out of the question for a 200-ton airplane. Furthermore, the large size of the 5,000-ton flying ship is a help in minimizing radiation exposure to passengers and crew. The 5,000-ton flying ship would be propelled by ducted fan propellers located in tunnels through the thick wings. Combinations of pusher and tractor propellers could be turned by electric motors with current from a central atomic power plant. With a 1,200-foot wingspread, a thirty-foot thickness would not be out of proportion, and tunnels expanded to a diameter of 20 feet at the propeller ducts could be accommodated. Ten such tunnels in each 600-foot wing would provide power for flight and be ideal for boundary layer and/or laminar flow control at the surfaces of the wings. A combination of hydroplane steps and of air bearings under the hull would facilitate take-off of the huge craft, although no runway length would limit the take-off run in open water. Perhaps the tourist passengers could accept a 400-knot air speed at moderate altitudes in return for all the luxuries of an ocean liner—staterooms, restaurant, theatre, shops, deck chairs and even a swimming pool. A thousand tons of express cargo and mail could be profitably carried.

For all-out high-speed travel, rocket-powered aerospace transports will make no two points on earth farther than two hours apart. With liquid hydrogen and oxygen propellant in controlled thrust motors, the transport could take off as a jet airliner does today, but would quickly go into ballistic mode and rocket up to 80 or 90 miles at 8,000 miles an hour. It would then shut off power and glide in ballistic trajectory to its destination, accomplishing a powerful reverse thrust re-entry followed by normal landing at aerodynamic flight speeds after deploying retractable wings. It might still be possible to make better time between New York and Paris than the limousine does in getting to the city from either airport.

Bionics and cybernetics are two favorite catchwords in the argot of technology. Bionics means the study of living functions as a way of making short cuts in technical applications. Fish and aquatic mammals that use echo returns similar to sonar, migrating birds that navigate by reference to the earth's magnetic fields, insect mosaic eye vision applied to the design of a battery of photo-electric eyes to measure aircraft ground speed are illustrations of the significance of bionics. Cybernetics is the science of automation. Machines can be guided through complex precision operations by command of a tape recorder, playing back through electronic controls. For example, a plywood plant in Tennessee has twenty-seven employees, mostly maintaining the machines, to make plywood by automated machinery, which would require twelve hundred assembly-line workers without automation.

A further step by combined application of bionics and cybernetics can be foreseen. This is the man-machine combination. To avoid exposure to radiation hazards, remote manipulation devices have been developed which permit the motions of the operator to be faithfully transferred to an artificial hand and arm on the other side of a shielding panel. When the operator picks up a dummy test tube, a real test tube of radioactive chemicals is picked up 25 feet away by his mechanical mimic; to pour its contents accurately into a flask requires no little practice on the part of the operator. Nevertheless, an astounding variety and complexity of manual remote-control functions can be done with this pantograph system of remote manipulation.

A more rational approach involves the use of a light, plastic armor suit completely enclosing the man, so that any movement he makes with any part of his body moves the suit correspondingly. The movements of the suit actuate electronic circuits which can translate them to corresponding movements of hydraulically actuated homologs of the man's muscles. For example, a man wearing the control suit could climb into an air-conditioned steel replica of himself, several times his size, and with no exertion, go through the motions of picking up and tossing aside one-ton boulders while the hydraulic muscles, cued by his movements, did the actual work. In another application, he could wear the control suit while looking at a three-dimensional television image projected from a closed-circuit TV camera in the head and eyes of his steel replica. The replica could be walked into a reactor control room in a disaster and, under direct visual control by the operator, could perform any required task to remedy a malfunction or avert a catastrophe. Firemen would no longer need to enter burning buildings, even though this might take the romance out of the rescue for a damsel in distress. Microphone ears for such a robot would be used in situations calling for hearing, and pressure and temperature transducers could communicate sensation to the human operator's fingers.

A more fantastic application would be to put such anthropomimic robots in a space station, slaved to human controllers in a ground-based duplicate of the station. The earthonaut and his mechanical astronaut twin would lead identical lives, except that the life-support requirements would be eliminated for the mechanical astronaut, and the human-risk element would be eliminated from the flight. A lunar station could be built, tested, and provisioned by such robot astronauts before sending a human to make direct astronomical and geophysical observations. Space exploration could be expedited and great savings could be realized by taking short cuts with a robot that would be unthinkable where human lives might be at stake. Certainly, the technical difficulties of reproducing the movements of the fifteen finger articulations and actuating them identically with those of a living hand are no more formidable than those of projecting human life-support requirements for a lunar voyage, especially when it is understood that robots will make one-way trips.

In our tradition, machines are expendable while men are not; and it is, therefore, in their possible military application that such robots may prove of greatest value. As expendable pawns in tactical experiments in war games, the robots may usefully demonstrate that every clever tactic employed in battle can be trumped. It is fondly hoped that the war-games strategists will be so bemused with the compliant robots that war will be relegated to these machines and their masters, leaving the rest of mankind to a better fate. Thus, automation, which now appears a threat to the economic livelihood of man, may be the means of using machines to take over the hazards of man's most odious and useless activity, the waging of war.

Indeed, machines which speed communication already have reduced the possibility of war. It is generally agreed that the human race resorts to war from lack of understanding that results from failure of communication. As means of communication have improved in rapidity and effectiveness, information has been transmitted with increasing clarity and detail about crucial events, with the result that national leaders talk and arbitrate their way out of crises that formerly would have been cause for war.

Electronic technology also has made outstanding contributions to news reporting with the inventions of the video-tape television camera and orbital communications satellites. Universal audio and visual reporting on national and international affairs has advanced the democratic process by keeping the public fully informed. No tyranny can long survive the power of informed public opinion.

Pictures, however, are but an incomplete medium of universal communication, even as words often are. Diversity of languages still presents the most formidable barrier to free exchange of information, still throws up obstacles of delay and distortion in the translation; and efforts to invent universal languages such as Esperanto only add a new one to the inventory. Electronic technology offers the best hope of a way out of this dilemma. Computers with vast memory drums and instant recognition of an image or a sound will be able to translate both written and spoken languages instantly, matching them with counterparts in other languages, sorting words into proper grammatical order and presenting the result in sound or in writing in the chosen idiom. The problem is formidable, but progress is being made. Here the compactness of solid-state electronic components, which amount to miniaturizing of transistors and printed circuits, will prove ideal for handling the complexity of circuits with the reliability and sensitivity required in translating machines. A billion solid-state components can be packed into a cubic foot of space. The idea of a wrist radio, considered science fiction ten years ago, gives way to the serious consideration of a television set in an engagement ring. Tomorrow may see universal long-distance, individual, video-audio communication via orbital satellite relay, with the words instantly translated into the appropriate language for the listener. The last great barrier to human understanding will be electronically erased, and with it all apartheid of human thoughts relegated to language chauvinism.

Behind the language barrier lies the prejudice barrier, secure only when entrenched in human ignorance. The war on ignorance has just begun. No illiterate with access to a television set broadcasting in his native tongue can ever be as ignorant as his predecessors who were without electronic communications. Education begins with learning how to communicate in written and spoken languages, through visual images and music, and in mathematical symbols. It continues as a process of gathering and organizing information for recall when it is needed. The next stage is learning how to rationalize information for solving problems and for application to real situations. Education becomes a creative function when information is projected into invention, discovery, the synthesis of new knowledge, the exploration of the unknown, the formulation of ideas and concepts. To be educated is to have learned how to learn; to have learned how to communicate learning; to have learned how to translate learning into action; to have learned how to create learning; and by all these means, to have learned how to seek and understand truth.

The ever increasing avalanche of knowledge from the contemporary technological revolution would be overwhelming were it not for help coming from the application of technology to teaching and learning techniques. The reiterative and rote memory aspects of learning can be accomplished with teaching machines that make auditory and visual presentation of material, put the student through training exercises and tests, and spare human teachers from drudgery in order to use their talents in the intellectual functions of teaching that are beyond the scope of teaching machines. The student is spared lost motion and is permitted to set his own learning pace. The method of presentation will be designed to maintain maximum interest and learning effectiveness. The time of day, most efficient work-rest cycles, the internal chemistry of the student, and the conditions of his classroom environment are factors that can be rendered ideal for rapid, effective learning.

Subjects requiring logic and problem solving will utilize computers with memory storage of entire courses of study, and teachers will be able to concentrate on the more creative levels of learning. The students selected and prepared by the machine and computer phases of teaching will be so well grounded that no time will be wasted in reviewing factual material, thus allowing the teacher to emphasize the development of ideas, and, in technical material, application and experimentation.

Graduate study and research can be facilitated by two other functions of computers: information retrieval, whereby storage memories can supply bibliographies or separate references, statistical tables, formulas for computations or any other desired information instantly from a constantly updated memory library; and operations research to evaluate statistically the factors of a research problem and to determine the most feasible plan for investigating it.

Throughout the educational process, the elimination of traditional, irrelevant, erroneous, or merely nonsensical material will be necessary in order to make room for factual and essential information which already approaches the saturation point in high-school and college curricula. Pedantic intellectual gymnastics have no place in a world where mastering a working knowledge of a technical field already taxes capable students to their limits. Eventually, this working knowledge will be a partnership between the scholar and his memory, and computer mechanical aids, in order to relegate to the machines those functions which they can perform for him, leaving his mind free to concentrate on aspects of knowledge with which machines cannot cope. With ever better tools, the effectiveness of human effort will be correspondingly amplified. This in turn will place greater emphasis on better minds, and on better human attributes demanded by many special requirements in tomorrow's world of sophisticated technology.

Having conquered his environment and equipped himself with technical aids that stretch his capabilities, man is confronted with the most formidable challenge of all—how to make a better man. The evolutionary and genetic approach cannot keep pace with changing requirements for ideal human specifications. Indeed, accelerating the evolution of man becomes increasingly important lest he be overtaxed in adapting himself to the environment created by his own inventions. The test-tube children of Aldous Huxley's *Brave New World* cease to be an amusing satire as modern biochemistry aided by electron microscopes unravels the chemistry of genes and the architecture of protein molecules. Their modification by radiant energies, the actions of enzymes and catalysts, may yet permit the control of germ-cell composition and the direct modification of the resultant human produced from a controlled beginning. Insidious effects of disease and chemical noxious agents can be eliminated during gestation and subsequent growth to maturity. This growth and this maturity can be controlled with ideal nutrition, hormone levels and programmed living activities to maintain health, youth, and effectiveness at an optimum far beyond the normal life expectancy of today. The elimination of degenerative disease and the prolongation of normal optimum anatomy and function of human organs will crown this creative achievement. A superior being living at physiological and mental age of 40 to a chronological age of 160 will be man's most magnificent scientific achievement. A sufficient number of individuals will provide the leadership to accelerate enormously the technical revolution and the future outlook of the human race.

Mankind groped and fumbled with flint and bone tools for bare survival through twenty-five millennia of the European Ice Age, stumbled by trial and error to the dawn of civilization in ten more millennia, walked with increasing purpose and effectiveness for fifteen centuries to the Renaissance, then found the key to five centuries of unlimited progress in the scientific method. This great legacy of technical achievement will enable man to better himself by conquering all the adversities of his environment and exploring the universe beyond the confines of his native planet, provided he does not fall victim to the thirty-minute war followed by the everlasting peace which could be unleashed from a hundred missile sites. Either way, the future of man is unlimited.